Chemical Measurements in Biological Systems

TECHNIQUES IN ANALYTICAL CHEMISTRY SERIES

Chemical Measurements in Biological Systems

Kent K. Stewart   Richard E. Ebel

WILEY-
INTERSCIENCE

A JOHN WILEY & SONS, INC., PUBLICATION

New York · Chichester · Weinheim · Brisbane · Singapore · Toronto

This book is printed on acid-free paper. ∞

Copyright © 2000 by John Wiley & Sons, Inc. All rights reserved.

Published simultaneously in Canada.

For ordering and customer service call. 1-800-CALL-WILEY.

Library of Congress Cataloging-in-Publication Data:

Stewart, Kent K.
  Chemical measurement in biological systems/Kent K. Stewart, Richard E. Ebel.
    p. cm. - (Techniques in analytical chemistry series)
  Includes bibliographical references and index.
  ISBN 0-471-13903-3 (cloth:alk. paper)
  1. Analytical biochemistry. I. Ebel, Richard E. II. Title. III. Series.

QP519.7 .S74 2000
572˙.36–dc21   99-045538

Printed in the United States of America.

10 9 8 7 6 5 4 3 2 1

We dedicate this book to our wives, Peggy and Edie; our children, Elizabeth, Cynthia, Richard, and Robert; and Ed and Beth. Also to our grandchildren, Erin, Laura, David, and Paige; and Alica and Wesley.

Contents

Foreword

Titles in the *Techniques in Analytical Chemistry Series* address current techniques in general use by analytical laboratories. The series intends to serve a broad audience of both practitioners and clients of chemical analysis. This audience includes not only analytical chemists but also professionals in other areas of chemistry and in other disciplines relying on information derived from chemical analysis. Thus the series should be useful to both laboratory and management personnel.

Written for readers with varying levels of education and laboratory expertise, titles in the series do not presume prior knowledge of the subject, and guide the reader step by step through each technique. Numerous applications and diagrams emphasize a practical, applied approach to chemical analysis.

The specific objectives of the series are:

- To provide the reader with overviews of methods of analysis that include a basic introduction to principles but emphasize such practical issues as technique selection, sample preparation, measurement procedures, data analysis, quality control, and quality assurance
- To give the reader a sense of the capabilities and limitations of each technique and a feel for its applicability to specific problems
- To cover the wide range of useful techniques, from mature ones to newer methods that are coming into common use

- To communicate practical information in a readable, comprehensible style; readers from the technician through the Ph.D. scientist or laboratory manager should come away with ease and confidence about the use of the techniques

Forthcoming books in the *Techniques in Analytical Chemistry Series* will cover a variety of techniques, including capillary electrophoresis, chemometric methods, biosensors, supercritical fluid extraction, surface and interface analysis, measurements in biological systems, inductively coupled plasma-mass spectrometry, gas chromatography–mass spectrometry, Fourier transform infrared spectroscopy, and other significant topics. The editors welcome your comments and suggestions regarding current and future titles, and hope you find the series useful.

FRANK A. SETTLE
Lexington, VA

Preface

The life sciences are in the midst of a revolution brought about by rapid advances in biochemistry, molecular biology, and biotechnology. Biochemistry is one of the cornerstones of this revolution. Today almost everyone involved in the life sciences, be they in teaching, industry, or the government, needs a basic foundation in biochemistry. A key component of this foundation is the understanding of the processes of making the chemical measurements in biological systems. We have tried to provide a book that would be an introduction to the fundamental concepts and principles of making chemical measurements in biological systems. Our goal is to provide an introduction for those in the life sciences on how to make chemical measurements in biological systems and how to evaluate and validate the data from those measurements. We wanted to provide some tools that those in the life sciences can use to answer the questions that are important to their disciplines.

The authors know of no book for those in the life sciences that is designed to teach the fundamentals of making chemical measurements in biological systems. While there are many analytical chemistry texts that emphasize gravimetric, titrimetric, and electrochemical analysis, these techniques are rarely used in the modern life science laboratory. The general emphasis of traditional analytical chemistry texts seems to be the assay of inorganic analytes and/or of the relatively simple matrices found in manufactured chemical products. There is little discussion of the assay of biological samples. There are several excellent texts for instrumental analysis, but these texts emphasize the instrumentation rather than the assay. Such texts are useful for those in analytical chemistry, but less useful for those in the life sciences. There are several texts for technicians that have good coverage of traditional assay methods; however, the approach of most of these texts is method and/or instrument specific. Such texts frequently lack a good discussion of the complexity of biological systems and the basic principles of making chemical measurements in biological systems. Discussions of the basic principles of making enzyme measurements and of using enzymes as reagents are often incomplete or totally missing in all but biochemistry laboratory textbooks. Similarly, immunochemical assays are only rarely discussed in analytical chemistry texts. A discussion of the need for assay validation and assay quality control is quite rare, and is usually only found in specialized books on those topics.

This book is designed to be used by advanced undergraduates, graduate students, and professionals in the life sciences such as biology, biotechnology, microbiology, the plant sciences, the animal sciences, genetics, human and animal nutrition, food science, and the ecological sciences. Most analytical chemistry books are addressed to those working in the traditional chemistry industry; this book is addressed to those working in the life science and biotechnology industries. While we believe that this book could prove useful to biochemistry majors and/or analytical chemistry majors, it is not specifically intended for those disciplines.

The basic idea behind the development of *Chemical Measurements in Biological Systems* is an emphasis on teaching fundamental concepts and principles of the measurement process, not on the instrumentation used to make the measurements. Specific methodologies are not emphasized. There are so many techniques and they are sufficiently complex that it is unrealistic to expect any analyst to know all the intricate details of each and every technique. However, it is important that analysts understand the fundamental concepts of chemical measurements in biological systems so that they understand how to select an appropriate analytic method from the vast array of existing methods, appreciate its usefulness and limitations, can modify an existing method to make it appropriate for the task at hand, and can critically evaluate the results of a given assay in a given matrix. Most important, analysts working in the life sciences today need to understand and be able to use multistep "problem-solving techniques."

The underlying problems in the chemical assay of analytes in biological matrices are the complexity of the matrices and the low concentrations of the analytes. Thus the topics of blanks, controls, and assay validation are emphasized. The goal of this book is to give students a sufficient background so that they can work in any laboratory using biochemical assays. The idea is for students to be able to read and understand a method description, to set up and run a given assay, to evaluate the data from a given analysis, and to present the results of analyses in a complete and concise form.

The topics of this book include those techniques that are in common use for chemical assays in biological systems. Those include pH, buffers, spectrophotometry, colorimetric reactions, measurement of enzyme concentrations, measurement of enzyme parameters including V_{max}, K, and K_is, the use of enzymes as reagents, basic chromatography, including TLC, LC, HPLC and GLC, electrophoresis, enzyme-linked immunosorbent assays (ELISA), data validation, and assay quality control.

This book does not cover the techniques of structural elucidation of small- and medium-molecular-weight natural products, since we believe that these topics have been well covered by others. We have chosen not to discuss a number of important but specialized topics such as capillary electrophoresis, GC-MS, LC-MS, SFC, SFE NMR, fluorescence spectroscopy, atomic spectroscopy, Northern and Southern blots, cell transformation, restriction mapping, cloning, DNA sequencing, DNA finger-printing, and radioisotopes. We believe that while these topics are important to many in the life sciences, they are generally used only in specialized laboratories and are not appropriate for an introductory book. However, we believe that the basic concepts emphasized in our book are equally important for these measurement systems.

We have assumed that the majority of our readers have access to and use computers routinely in the laboratory environment. The computations presented in this book are much easier to do if the reader uses spreadsheets such as Graphic Analysis for Windows^{TM}, Microsoft^{R} Excel, or Corel^{R} Quattro-Pro. While manual computations may also be used, it is our belief that such manual manipulations can be tedious, and we strongly recommend that the reader utilize the power of spreadsheets for data manipulation and presentation.

Given the current state of the analytical chemistry texts for those in the life sciences, the authors believe that the selection of the topics for this book is unique. We hope that this text will fill a gap that currently exists in the education of those in the life sciences who need to make chemical measurements in biological systems.

Kent K. Stewart
Richard E. Ebel

Acknowledgments

We have developed the concepts for this book over many years of discussion with many colleagues too numerous to mention. However, special achnowledgement to both our colleagues at Virginia Polytechnic Institute and State University, and Kent Stewart's colleagues at the Food Composition Laboratory, Agricultural Research Service, United States Department of Agriculture. We also want to give special acknowledgement to our teachers, Dr. Lyman Craig and the instructors at University of California at Berkeley, Florida State Universuty (KKS), and the instructors at Northwestern University and University of Wisconsin (REE). Rita Wilkinson gave very valuable assistance in preparing many of the figures. We thank all of these people.

Kent K. Stewart
Richard E. Ebel

1. Introduction

Measurement: "When you can measure that you are talking about, and express it in numbers, you know something about it; but when you cannot express it in numbers, your knowledge is of a meager and unsatisfactory kind; it may be the beginning of knowledge, but you have scarcely, in your thoughts, advanced to the stage of SCIENCE."

—*Lord Kelvin*

Chemical measurements in biological systems are a balance between the theoretical and the pragmatic. If we glorify the theory because it is pure and demean the pragmatic because it is too common, then it is unlikely that we will be able to make meaningful measurements in biological systems; AND if we ignore the theory and are only pragmatic, then most assuredly we will end up with incorrect interpretations of our analytical data.

INTRODUCTION

The life sciences are in the midst of a revolution brought about by rapid advances in biochemistry, molecular biology, and biotechnology. Today almost everyone involved in the life sciences, be they in teaching, industry, or the government, needs a basic understanding of the processes of making chemical measurements in biological systems. Today analysts have numerous techniques to choose from when they wish to assay for a given component. The question no longer is "What

1

is the method for the assay of a given component in a sample?"; rather the question is "Which of the several available techniques should be used to assay for this component in this sample?" There are so many techniques and they are sufficiently complex that it is unrealistic to expect any analyst to know all there is to know about each analytic technique. While it is not as critical that analysts understand the intricate details of each and every technique, it is important that they understand the fundamental concepts of chemical measurements in biological systems so that they can make intelligent choices between techniques, can properly evaluate the use of individual techniques, and can critically evaluate the results of given analyses.

Definition: An assay is a method for the determination of the level of a given analyte in a given sample or set of samples.

Definition: An analysis is a determination of the level of a given analyte in a given sample or set of samples.

Definition: An analyte is the substance (element, ion, or compound) being analyzed.

Definition: The analyte matrix is the chemical environment of the sample that contains the analyte.

Biochemical chemical analyses are done to answer specific questions. Typical questions[1] being addressed by analysts doing chemical assays in biological systems include: Is the analyte in the sample? What is the level of the analyte in the sample? Does this sample have a concentration of the analyte higher than (lower than, equal to) that sample? Does the product meet the specification for the levels of the analyte? What are the identities of the individual components of a class of compounds (e.g., individual amino acids in a protein)? Does the treatment alter the sample? What is the mean level of the analyte in the population represented by the samples? What is the statistical distribution of the analyte in the population represented by the samples?

Biological Activity

Analysts who make chemical measurements in biological systems have some special challenges. When an analyst is asked to assay for a biological activity, defining the question can become complex. Those analysts who make chemical measurements in biological systems need to understand that they will frequently

---

[1] The issues related to determining the structure of a given compound have been well covered elsewhere and will not be discussed in this book.

be working at the interface of biology and chemistry. Biology is concerned with stimulus and response of living things. Chemistry is concerned with chemical structures and reactions at the molecular level. Those analysts making chemical measurements in biological samples often deal with two different scientific cultures, each with their own perspectives. Seemingly identical terms mean different things in the two areas of science. Furthermore, there are a number of cases where a single biological response is invoked by a number of chemical compounds. Also, there are numerous cases where multiple biological responses are invoked by a single chemical compound. For example, the concept of "vitamin" is a biological activity. There are six primary chemical compounds that have vitamin B_6 activity and more than 20 that have vitamin A activity. To do a chemical assay for vitamin B_6 activity, one would have to assay for six different chemical compounds; for vitamin A activity, for about 20 chemical compounds.

Concentrations

The range of concentrations of chemical compounds that elicit biological responses is quite large. The concentration of water in many biological systems approaches 55 molar, and many life processes cease when the biological matrices reach water levels where the water activity[2] is less than 0.2. Lysergic acid diamide can elicit quite dramatic effects in humans at a concentration of 1 microgram per 70 kg: a concentration of about one part in 100 billion. There are numerous cases in which a nanoMolar concentration of a particular compound will elicit a biological response. Given that chemical assays at such low concentrations can be difficult, one might think that the simple solution is to do biological measurements for the biological activity of these analytes. However, experience has taught us that biological measurements are very complicated. Biological responses are frequently modified by the concentrations of compounds other than the analyte being measured. Biological measurements can be highly variable. Large numbers of assays are frequently needed to obtain the desired assay precision. These large numbers of assays almost always lead to a high cost. Many biological responses occur only slowly over time, and thus biological assays frequently have long turnaround times. Biological responses are often species specific, and the results obtained with one species do not necessarily duplicate the results that would be obtained in another species. Thus biological measurements, as such, are usually done only when there are compelling reasons to do so. Chemical and physical methods are the usual methods of choice for the determination of the concentrations of components that are defined by their biological activities. Still, there are a number of analytes where no effective chemical or physical assays have been developed that yield satisfactory results at the concentrations that elicit the biological activity, and biological assays are the only methods available (e.g., the assay of active botulism toxin is still best done by the mouse toxicity assay).

---

[2] See Chapter 2 for a more detailed discussion of water activity.

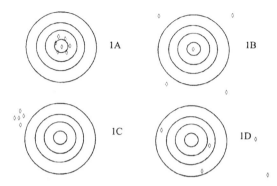

Fig. 1.1. Accuracy and precision: (A) Accurate and precise; (B) accurate, but imprecise; (C) inaccurate but precise; (D) inaccurate and imprecise.

Accuracy and Precision

Two of the desired characteristics of any assay for analytes in biological systems are accuracy and precision. The concepts of accuracy and precision are illustrated in Figure 1.1.

Definition: The accuracy of a result is how close it is to the "true" answer.

Definition: The precision of a result is its repeatability (reproducibility).

An accurate assay method is one that has the selectivity to yield results that are due only to the analyte. An inaccurate assay method is one that yields results that are (at least partially) due to nonanalytes. Inaccuracies are often systematic in nature and have the characteristic that they are often not random. Systematic errors (biases) are introduced in a measurement system through such factors as contaminants, drift, improper calibration and calculation, and report generation errors. The usual result of the presence of interfering compounds is to obtain values that are too high. Less frequent, but still common, are the low values obtained when some compound inhibits the response; an example would be the microbiological assay of vitamers[3] in foods that have appreciable concentrations of antimicrobial activity. Special problems of inaccurate answers can result from working at very low signal-to-noise ratios.

Determining the accuracy for the assay can be very difficult, and it is current dogma that while the accuracy of an assay can be disproved, it cannot be proven. Still there are a number of actions that the analyst can take to improve the probability that the assay results are accurate. At a minimum, the analyst should

---

[3] The term *vitamers* is defined as that group of chemical compounds that elicit the biological activity of a vitamin.

determine that the assay used will give a proportional response to changes in the concentration of the analyte within the concentration range of interest. The analysis of reference samples of known concentration along with the assay of the sample with unknown concentrations is frequently done in the attempt to validate the accuracy of the assay.

The appropriate blanks and controls should be run to demonstrate that the other components in the matrix neither give a false positive response that might be taken as a response due to the analyte nor somehow interfere with the response of the assay system to the analyte and give a low value for the level of the analyte. Many times a known amount of the analyte will be added to the sample, and recovery studies will be done to demonstrate the accuracy of the original results. The concepts of the verification of assay accuracy will be discussed in more detail in the remainder of this book. A brief introduction to statistical tests that can be used to identify possible biases between assay methodologies is given in Appendix 2.

Precision

All measurements have variability or imprecision. The precision of an assay is often dependent upon the natural variation of the samples, the skill of the analyst, the assay method being used, the variability of the analytical instrumentation, the variability of the reagents used for the assay, the variability of the reaction temperature and time, and the analyte concentration. The usual goal is to have the measurement methodology be precise enough so that the real differences between samples can be determined. The precision of a result can be determined by doing repeated assays of the same sample for the same analyte. Knowledge of the variance of an assay is an important part of the ability of an analyst to demonstrate that the assay will answer the question being asked. A brief introduction to the precision and the statistical components of chemical measurements in biological systems, and their impact on the answering of biological questions is given in Appendix 2.

The Analyte

The chemical structure and physical properties of the analyte are the determining factors of what measurement techniques might be used to determine the level of the analyte in a solution. For example, proteins have the chemical features listed in Table 1.1, many of which might be used for their analysis. Any physical or chemical technique (or a combination of techniques) that responds to changes in the concentrations of any of these chemical features could be used to determine the concentration of a pure protein in water. Unfortunately, there are very few situations requiring chemical measurements in biological systems where the analyte is found in pure form as a solid or in solution. Many components of biological matrices interfere with assays for most of the physical and chemical

TABLE 1.1.   Chemical Features of Proteins

Peptide bonds
Amino groups
Carboxyl groups
Aromatic groups
    Phenyl
    Indole
Imino groups
Alcohol groups
Phenol groups
Sulfhydryl groups
Imidazole groups
Hydrophobic areas
Enzymatic activity
Immunological sites
Bound chromophores
Other bound chemical compounds (e.g., metals)
Size
Shape
Charge

properties of analytes at their normal concentrations. Therein lies the challenge for the analyst.

Biological Matrices

While the chemical structure and physical properties of the analyte determine which measurement techniques may be used, properties of the matrix and the concentration of the analyte in that matrix determine which measurement techniques may not be used to assay the analyte in the matrix (Fig. 1.2). Biological matrices are chemically complex. For example, the chemical compositions of typical cells are shown in Table 1.2, and typical numbers of compounds per chemical class normally found in cells are shown in Table 1.3. Metabolite concentrations in most cells range from 100 mM to 1 nM.

As a specific example of the selection of an assay method, examine the analysis of protein in blood plasma, a common biological matrix of about average complexity. Blood consists of several families of cells and formed elements[4] in a protein-rich solution called the *plasma*. The cellular and formed element components, which account for about 45% of the total blood volume, can be removed with centrifugation and the plasma isolated. A typical composition of blood plasma is shown in Table 1.4. Note that there are a large number of

[4] These families of cells and formed elements include the erythrocytes (about 5 million per mL), the leukocytes (about 5000 per mL), and the platelets (about 300,000 per mL).

TABLE 1.2. Approximate Chemical Composition of a Typical Mammalian Cell[a]

	Percent of Total Cell Weight
Component	Mammalian Cell
H_2O	70
Inorganic ions (Na^+, K^+, Mg^{2+}, Ca^{2+}, Cl^-, etc.)	1
Miscellaneous small metabolites	3
Proteins	18
RNA	1.1
DNA	0.25
Phospholipids	3
Other lipids	2
Polysaccharides	2

[a]Reference B. Alberts, D. Bray, J. Lewis, M. Raft, K. Roberts, J.D. Watson, 1989. *The Molecular Biology of the Cell,* 3rd Ed., Garland Pub., NY, p. 88.

components at different concentrations and the presence of other compounds in the plasma that have many of the same functional groups that were listed for proteins in Table 1.3 (i.e., amino groups, carboxyl groups, aromatic groups, thiol groups, and chromophores). Thus the use of assays for these groups as assays for proteins would give false values for the protein concentration in plasma. Likewise, the common analytical detection techniques of weight, refractive index, electrical conductivity, infrared spectroscopy, mass spectrometry, electrolytic methods (anodic, voltammetry, polarography), ion-selective electrodes, atomic absorption, atomic emission, NMR, fluorescence, chemiluminescence, Raman spectroscopy, and electron spin resonance are not particularly useful for the measurement of proteins in the blood plasma because many other components of

TABLE 1.3. Approximate Numbers of Compounds in a Living Organism by Chemical Category

Category	Approximate Numbers
Amino acids, precursors, and derivatives	120
Nucleotides, precursors, and derivatives	100
Fatty acids, precursors, and derivatives	100
Carbohydrates, precursors, and derivatives	250
Vitamins, coenzymes, isoprenoid, and porphyrin	300
Enzymes and other proteins	5000

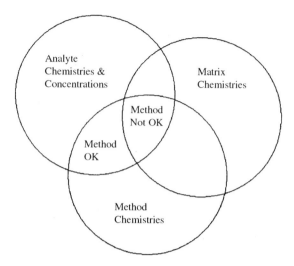

Fig. 1.2.   Selection of appropriate assay methodologies.

plasma would give positive responses. Most assays for total protein in plasma use uv/vis spectroscopy coupled with colorimetric assays for peptide bonds or hydrophobic regions. Assays for specific proteins can often be done by measuring the specific enzymatic activity of a protein, if there is one, or measuring the concentration of the specific immunological sites of those proteins.

Analytical Process

The analytical process consists of three major steps (Table 1.5): defining the problem, making the measurements, and reporting the results. The measurement process consists of the actual sampling, sample pretreatment, measurement, result computation, and result validation. The measurement process is a holistic one in which the actions at any one of the steps will influence the results of later steps. Thus the analyst must understand the impact of the choice of techniques at each step of the measurement process on the results of the later steps in the measurement process. An error at any stage of the measurement process will almost always lead to an incorrect result! Given the underlying problems in the assay of analytes in biological samples of the complexity of the matrices and the frequent low concentrations of the analyte, the analyst needs to pay particular attention to the chemistry of the analyte and of the matrix when selecting the components of an assay method. A wise analyst must assume that unexpected interferences may exist in the biological matrix and plan accordingly.

Table 1.4. Components of Plasma

Component	Normal Concentration (Range)
Water	94% v/v
Sodium	3.1–3.3 g/L
Potassium	137–196 mg/L
Calcium	84–104 mg/L
Chloride	3.5–3.8 g/L
Bicarbonate	1.3–1.6 g/L
α-Amino nitrogen	30–55 mg/L[a]
Ammonia	0.80–1.10 mg/L
Glucose (fasting)	0.7–1.1 g/L
Hormones	
ACTH	15–70 ng/L
Aldosterone	50–500 pg/mL
Cortisol	50–250 μg/mL
11-Deoxycortisol	>75μg/L
Growth hormone	
Children	>10μg/L
Adults	<5 μG/L
Insulin	43–187 pM
Prolactin	0.08–6 nM
Total tri-iodothyronine	0.7–1.9 μg/L
Total thyroxine	40–120 μg/L
Testosterone (males)	30–110 μg/L
Testosterone (females)	0.25–0.9 μg/L
Lipid components	
Cholesterol	1.2–2.2 g/L
Cholesterol esters	60–75% of total cholesterol
Phospholipids	90–160 mg/L as lipid phosphorus
Total fatty acids	1.9–4.2 g/L
Total lipids	4.5–10 g/L
Triglycerides	0.4–1.5 g/L
Metals	
Copper (all forms)	1–2 mg/L
Iron (females[b])	0.5–1.5 mg/L
Lead	<500 μg/l
Phosphorus (inorganic)	30–45 mg/L
Sulfate	5–15 mg/L
Organic acids	
Lactic acid	0.6–1.8 meq/L
Plasma proteins	60–84 g/L
Albumin	35–50 g/L
Ceruloplasmin	270–370 mg/L
Globulin	23–35 g/L

(*continued*)

Table 1.4. (*continued*)

Component	Normal Concentration (Range)
Enzymes[c]	
Aldolase	1.3–8.2 mU^d/mL
Amylase	4–25 U/mL
Creatine phosphokinase	
Female	5–35 mU/mL
Males	5–55 mU/mL
Lactic dehydrogenase	60–120 U/mL
Lipase	2 U/mL
5′ Nucleotidase	0.3–3.2 Bodansky U/mL
Phosphatase (acid)	
Female	0.01–0.56 Sigma U/mL
Male	0.13–0.63 Sigma U/mL
Phosphatase (alkaline)	113–39 IU/mL
Infants and adolescents	up to 104 IU/mL
Transaminase	10–40 U/mL
Phenylalanine	up to 20 mg/L
Urea nitrogen	80–250 mg/L
Uric acid	30–70 mg/L
Vitamins and precursors	
Ascorbic acid	4–15 mg/L
Carotenoids	0.8–4.0 µg/mL
Vitamin A	0.15–0.6 µg/mL
Vitamin B_{12}	90–280 pg/mL
Bilirubin (total)	10 mg/L
Creatinine	6–15 mg/L

Source: R.E. Scully and B.U. McNeely, *New England Journal of Medicine* **302**(1), 37–48 (1980).

[a]The average molecular weight of the amino acids found in plasma is 100. Thus the approximate total weight of the free amino acids is about 100 times the α-amino nitrogen concentration.

[b]Higher in males.

[c]A unit of enzyme is normally defined as that amount of enzyme that will catalyze the formation of 1 µmole product per minute. If other terms accompany the unit designation, then these are specialized units. See the table reference for more details on such units.

[d]1 mU is 10^{-3} enzyme units.

Elements of Chemical Measurements

Many good chemical measurements in biological system are a result of the successful combination of the elements of detection, identification, and quantification. Good methods are based upon the accumulated knowledge of the chemistry of the analyte, knowledge of the chemistry of the nonanalyte

TABLE 1.5. The Analytical Process

1. Defining the problem
2. Making the measurements
   Sample selection
   Sample pretreatment
   Measurement
   Identify
   Quantify
   Computing the results
   Validating the results
3. Reporting the results

components of the sample matrix, and the characteristics of calibrated instrumentation for the separation, detection, and quantification of the analyte. Many biological samples have complex (and sometimes unpredictable) compositions, and that coupled with the normal problems associated with the analyses of the low analyte concentrations found in many biological samples strongly suggests that the potential for errors is rather significant. Thus wise analysts ensure that final analytical data sets are verified by the appropriate quality-control procedures.

Accurate chemical analyses require an understanding (from either refereed chemical literature or documented laboratory experiments) of both the known chemistry of the analyte and the known chemistry of the nonanalyte components of the matrix. Such known chemistries include the chemical structures of the analyte and nonanalyte components, the physical constants of each component [e.g., boiling point, melting point, $pK_a(s)$, spectra as determined by various spectroscopic techniques, oxidation/reduction potentials, solubilities in various solvents, etc.], the separation chemistries of each of the components, and the chemical reactions with other compounds, including the chemical structures of the reaction products, and the kinetics and equilibrium constants of the reactions. In most cases, no matter which process is used, the assay system will require calibration with a sample of known identity with a known concentration. Calibration is one of the key steps to ensuring that analytical processes are accurate.

Good analysts will routinely calibrate their instruments and chemical assay systems and document those calibrations. Really good calibrations permit the analyst to be able to trace the components of each analytical measurement result back to known and accepted units of measure and/or instrumentation and/or chemistries and/or reference material. We will discuss the use of calibrations and other techniques used to improve the accuracy of measurements throughout the book. While not all the above-mentioned chemical information is required for a good analysis, insufficient knowledge of the chemistry of an analyte or nonanalyte component may lead to inaccurate or imprecise results.

TABLE 1.6. Mechanisms for Determination of an Analyte Concentrations

The amount of analyte is proportional to:
1. The magnitude of the signal or
2. The number of signals or events or
3. The amount of reagent required to react completely with the analyte or
4. The rate of change in the concentration a compound

An appropriate separation process is one that quantitatively separates the analyte from the nonanalyte components of the sample matrix. Such an appropriate separation process will depend upon the chemistries of the analyte and the nonanalyte components and upon the chemistries of the analytical process itself. Detection of the analyte will depend upon the chemistry of the analyte, the detection chemistry (if any), and the chemistry and/or physics of the detector. Identification of the analyte will depend upon its behavior in the separation process, the reaction products of any utilized chemical/biochemical reactions, the biological responses utilized, and the physical constants of the analyte. Most identifications are made by a combination of these factors. Under most circumstances, the analyte must meet all the selection criteria being used for the identification. Failure to meet one of the criteria usually means that the compound being detected is not the analyte.

Quantification of an analyte takes place through one of the following four mechanisms: (See table 1.6) (1) The amount of analyte is proportional to the magnitude of the signal (or some mathematical function of the magnitude of the signal), for example, weight, light absorption (emission, etc.), voltage, resistance, current, etc. (2) The amount of analyte is proportional to a count of the number of signals or events (or some mathematical function of the number of signals or events), for example, molecule or atom counting, photon counting, and radioactive measurements. (3) The amount of analyte is proportional to the amount of the reagent required to react completely with the analyte, for example, all titrations and other stoichiometric reactions. (4) The amount of analyte is proportional to the rate of change in the amount (concentration) of a reaction product or of the original analyte (or some mathematical function of the rate of change of the signal), for example, all kinetic assays and chemiluminescence assays. The quantification of an analyte is based upon a comparison of the response of analytical process from a sample of unknown concentration with the response of that analytical process from a sample of known composition and concentration. Most, if not all, quantification processes have only a limited linear range that must be determined if accurate and precise concentrations are to be obtained. Verifications of the analytical results are done by concurrent analyses of pure analytes and known mixtures of known composition and concentration, and by the use of the appropriate blanks, controls, and external and internal standards.

A key factor[5] in the choice of assay methods is the capability accurately to measure the analyte in the matrix at the concentrations and with the precision that will answer the experimental question. That implies that an acceptable assay method will contain a series of steps that result in the measurement of some unique physical, chemical, or biological attributes of the analyte at the concentrations found in the biological matrix in the presence of the other components found in the matrix. In most cases an acceptable method will utilize a selective physical–chemical detection system (e.g., uv/vis spectrophotometry), and/or a selective chemical reaction (e.g., the biuret reaction), and/or a selective biochemical reaction (e.g., an enzyme reaction), and/or a selective biological reaction (e.g., immunological binding), and/or a selective separation technique (e.g., liquid chromatography). It is not infrequent that successful, modern assay methods use various combinations of these selective techniques to attain the necessary assay selectivity while measuring low concentrations of the analyte. At present most of the common chemical assays in biological systems utilize the detection techniques of uv/vis spectrophotometry, chemical fluorescence, isotopic counting techniques, and the pH electrode. These detection techniques are frequently combined with selective colorimetric and/or enzymatic reactions. The commonly used separation techniques include solvent extraction, analyte solubility, high-speed centrifugation, liquid chromatography (ion exchange, size exclusion, affinity, and reversed phase), gas chromatography, and electrophoresis. The assay of enzymatic and immunological activities is common. Primarily because of the problem of matrix interferences and the problems associated with the assay of low levels of analytes, the other common analytical techniques of weight, titrations, refractive index, electrical conductivity, infrared spectroscopy, electrolytic methods (anodic, voltammetry, polarography), ion-selective electrodes (other than the glass pH electrode), atomic emission, NMR, atomic fluorescence, chemiluminescence, Raman spectroscopy, and electron spin resonance are rarely used in the routine chemical assays of biological systems. Specialized laboratories use atomic absorption and flame emission spectrophotometry in the determination of the trace metal contents of biological systems; mass spectrometry as a detection system for gas or liquid chromatography; and refractive index, electrical conductivity, and voltammetry as detection systems with liquid chromatography. However, the use of these techniques is expensive, requires special training, and is rare in most life science laboratories.

The complexity of biological samples and the low levels of many analytes leads to frequent interference with the assays of individual analytes in biological

---

[5] There are several factors that go into method selection, and they are discussed in Chapter 11.

samples. Assay quality control is thus prudent. Given the potential for error, the validation of the individual data sets is strongly advised. There are several ways to validate data sets, including the use of standard methods, standard instruments, certified analysts, certified algorithms, standard reference materials, internal standards, the method of standard additions (spiking), audit trails, and pool samples. The basic concepts of data validation and assay quality control will be recurrent themes of this book and will be discussed in detail in Chapter 10.

The key components of chemical assays of analytes in biological systems are identification, quantification, and verification. In this book we will discuss the basic tools analysts use to make chemical measurements in biological systems. These tools include the use of pH buffers, spectrophotometry, colorimetric reactions, the measurement of enzymes and their parameters, the use of enzymes as reagents, the basic separation tools (e.g., extraction, chromatography, and electrophoresis), and enzyme-linked immunosorbent assays (ELISA). A common theme will be the use of those tools that can aid in the validation of results when the potential for matrix interference exists.

2. Water, pH, and Buffers

WATER

Water is crucial to life. Seventy percent or more of the weight of living organisms is attributable to water. Water is an extraordinary solvent. Many low-molecular-weight nutrients, metabolites, and metabolic products are dissolved in the intracellular water. Water is a primary source of nutrients and toxicants for biological organisms and is a significant vehicle for the disposal of metabolic wastes. Many chemical and biochemical reactions affecting living organisms occur in aqueous environments. Water dictates cell structure and is a primary determinant of most macromolecular structures, including proteins, RNA, DNA, and the complex carbohydrates. Given its influence on macromolecular structure, it should not be surprising that the level of the water content of the macromolecules is critical to their function.

Water Content

Knowledge of the water content is an important and key piece of information for the analysis of many types of biological samples. Since the water content of biological samples can vary significantly, the concentration of components in the aqueous portion of the sample can vary accordingly. Biological sample composition data are often reported on a dry weight basis to provide a constant reference point and reduce confusion. However, most analyses of biological samples are performed on "as is" samples, not on dried samples. The determination of the water content of these "as is" samples is made so that the results of the original analyses can be reported on a dry weight basis. Usually the water content of biological samples is found by drying the biological samples to constant weight at a given temperature and atmospheric pressure. There are a variety of methods for making this determination, including lyophilization and drying in a hot air or vacuum oven.

Water Activity

Some water molecules in a sample are tightly bound to components of the analyte and/or its matrix and are usually not available for chemical and biochemical reactions. Thus the chemical reactivity of the water in a sample is frequently not equal to its concentration. The chemically reactive water is defined as the percent of the total water that behaves as free water. This percentage of water that behaves as free water is defined as the water activity (a_w). The rates of many chemical reactions in biological systems decrease significantly as water activity decreases. Most organisms cannot grow below a water activity of 0.5, which explains why they regulate their water content within a very narrow range. Thus water activity (a_w) can greatly affect the stability of items such as foods. However, most biological compounds and even some viable organisms are stable in their dry state and can be stored for considerable periods at low a_w. There are several ways to measure a_w. The most common method is to measure the amount of water in the air that is in equilibrium with that sample (i.e., to measure the relative humidity, RH, of the air). Relative humidity is defined by Eq. 2.1 and the a_w by Eq. 2.2. Since a_w is a thermodynamic variable, the value of a_w varies with temperature (Eq. 2.3). Thus the temperature should be fixed when making a measurement of a_w and should be indicated whenever an a_w measurement is reported.

$$\%RH = \frac{p_{H_2O}}{p_{\text{pure } H_2O}} \times 100 \tag{2.1}$$

where p_{H_2O} is the partial vapor pressure of water in the system and $p_{\text{pure } H_2O}$ is the vapor pressure of pure water at the same temperature.

$$a_w = \frac{RH}{100} \tag{2.2}$$

$$\frac{d \ln a_w}{d(1/T)} = \frac{-\Delta H}{R} \tag{2.3}$$

Traditionally, relative humidity is measured from the increase in weight of dry cellulose equilibrated with the air above a biological sample (see Ref. 2.1). The

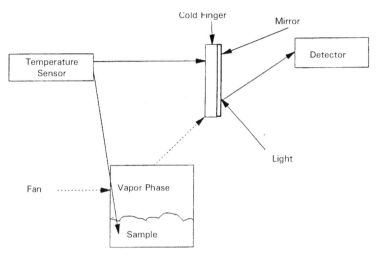

Fig. 2.1. A schematic of the CX-1 water analyzer.

water activity was then found by reference to a calibration curve prepared by equilibration of dry cellulose with the air above salt solutions of known water activity. While such measurements are time consuming and tedious, the measurement of the a_w could be accomplished using a simple apparatus and a laboratory balance. A more modern method of finding the relative humidity and thus a_w is based upon determination of the dew point of the air equilibrated above a biological sample. The dew point is the temperature at which the relative humidity of the air equals 100%. The measurement of the dew point is based upon the fact that the water-carrying capacity of air decreases with decreasing temperature. The relative humidity of an air sample can be measured by determining how much below the air temperature the temperature of a solid device needs to be to cause dew formation (condensation). This temperature difference depends on the original relative humidity. A schematic of an instrument used to measure relative humidity using this principle is shown in Figure 2.1.

Hydrogen Ions and Hydroxyl Ions

Water can dissociate, forming protons, hydrogen ions (H^+), and hydroxyl ions (^-OH) (Fig. 2.2).[1] Because of their small sizes and high charge densities, the H^+

---

[1] Protons (H^+) are hydrated and exist as H_3O^+ (hydrogen ions) or as $H^+(H_2O)_n$ in solution. For simplicity, we will use H^+ without reference to the hydration state. The term *protonated* refers to the form of a compound that has a bound, but dissociable, H^+, and unprotonated refers to the form of a compound that has lost a dissociable H^+.

$$H_2O \;\underset{\longleftarrow}{\overset{\longrightarrow}{}}\; H^+ + \;^-OH$$

Fig. 2.2. Dissociation of water to form protons (H^+) and hydroxyl ions (^-OH).

and ^-OH ions are extremely reactive, and alterations in their concentrations have profound effects on chemical and biological systems. Thus knowledge and control of the concentrations of protons and hydroxyl ions are important when making chemical measurements in biological systems.

pH

Hydrogen Ion Concentration and pH

The acidity of an aqueous solution is defined by the concentration of hydrogen ions $[H^+]$.[2] The higher the H^+ concentration, the more acidic is the solution. In aqueous solutions, $[H^+]$ usually has a concentration between ~ 1 and 1×10^{-14} M. Aqueous solutions with $[H^+] \sim 1$ to 0.99×10^{-7} M are considered acid, and those with $[H^+]$ 1.01×10^{-7} to 1×10^{-14} M are considered basic. Those with $[H^+]$ of 1.00×10^{-7} M are considered neutral. Traditional standard numerical notation for this large concentration range can be cumbersome, and errors are frequently made in writing and interpreting this type of notation. For ease of communication, $[H^+]$ is usually expressed in terms of pH. pH is defined as the negative logarithm of the molar $[H^+]$ or as the logarithm of the reciprocal of the molar $[H^+]$, which is equivalent (Eq. 2.4). For example, the pH of a 10^{-5} M solution of hydrogen ions is 5, while that of a 10^{-4} M solution is 4. Thus, as $[H^+]$ increases, the pH decreases.

$$pH = -\log_{10}([H^+]) \equiv \log_{10}\left(\frac{1}{[H^+]}\right) \tag{2.4}$$

Said in another way, the lower the pH, the higher the $[H^+]$, that is, the more acid in a solution. Given the range of $[H^+]$ presented, the pH range of aqueous solutions is from pH ~ 0 to ~ 14 ($[H^+] \approx 1$ to $\approx 1 \times 10^{-14}$ M). In pure water,[3] the pH is 7.0 and the $[H^+]$ is 1×10^{-7} M. The pH $+$ pOH always equals 14 in aqueous systems.

---

[2] While some authors have also dealt with $[^-OH]$ using pOH notation, we believe that the combination of pH and pOH can be confusing to the novice in the field, and we will restrict our usage to that of pH. If the reader needs to convert from pOH to pH, pH $= 14 -$ pOH in aqueous systems.

[3] Pure water is not commonly found in biological systems. Most frequently biological systems contain a variety of ionized components, including dissolved carbon dioxide (CO_2). In solution, CO_2 is hydrated to carbonic acid (H_2CO_3), which subsequently dissociates to yield H^+ and bicarbonate ion (HCO_3^-). Thus water exposed to air usually has a pH below 7.

Acids donate H⁺

Acid → H⁺ + conjugate base

Bases accept H⁺

Base + H⁺ → conjugate acid

Fig. 2.3. Brönsted-Lowry acids and bases.

Brönsted–Lowry Theory of Acids

The Brönsted–Lowry theory of acids and bases defines an acid as a proton donor and a base as a proton acceptor (Fig. 2.3). In other words, an acid yields hydrogen ions when dissolved in water, which increases the $[H^+]$ and decreases the pH of the solution. On the other hand, a base reacts with hydrogen ions from water when dissolved in water, decreasing the $[H^+]$ and increasing the pH of the solution. Removing hydrogen ions from water results in the production of ^-OH (Fig. 2.2). There are cases (Table 2.1) in which a compound can be either an acid or a base depending upon the reaction under consideration. For example, bicarbonate ion (HCO_3^-) is an acid if it donates a proton to form carbonate (CO_3^{2-}) but is a base if it accepts a proton to form carbonic acid (H_2CO_3).

Strong Acids and Bases

There are strong acids and weak acids. The tendency of a compound to donate or accept protons when it is dissolved in water depends upon the strength of the acid or base, respectively. A strong acid (e.g., HCl or HNO_3) is one that completely dissociates (ionizes) when dissolved in water at low concentrations to yield protons (H^+) and the conjugate base of the acid.

$$acid \longrightarrow H^+ + conjugate\ base$$
$$HCl \longrightarrow H^+ + CL^-$$

TABLE 2.1. Some Brönsted Acid and Base Pairs

Brönsted Acid		Conjugate Base	Acid Name
HCl	$\rightleftharpoons H^+$	$+Cl^-$	hydrochloric acid
CH_3COOH	$\rightleftharpoons H^+$	CH_3COO^-	acetic acid
H_2CO_3	$\rightleftharpoons H^+$	HCO_3^-	carbonic acid
HCO_3^-	$\rightleftharpoons H^+$	CO_3^{2-}	bicarbonate
$^+H_3NCH_2COOH$	$\rightleftharpoons H^+$	$^+H_3NCH_2COO^-$	glycine (fully protonated)
$^+H_3NCH_2COO^-$	$\rightleftharpoons H^+$	$H_2NCH_2COO^-$	glycine (zwitterion)
NH_4^+	$\rightleftharpoons H^+$	NH_3	ammonium ion

Examples of strong acids include hydrochloric acid, nitric acid, and the first proton of sulfuric acid. A strong base is one that completely dissociates when dissolved in water at low concentrations to yield a hydroxyl ion ($^-$OH) and its counter ion.[4]

$$\text{base} \longrightarrow {}^-\text{OH} + \text{conjugate base}$$
$$\text{NaOH} \longrightarrow {}^-\text{OH} + \text{Na}^+$$

Examples of strong bases are sodium hydroxide and potassium hydroxide. Weak acids and bases are only partially dissociated when dissolved in water. Examples of weak acids include acetic acid, all the protons of citric acid, the second proton of sulfuric, and all the protons of phosphoric acid. An example of a weak base is aqueous ammonia (ammonium hydroxide).

With complete dissociation of a strong acid, the concentration of protons and the conjugate base of the acid are a function of the initial concentration of the strong acid. The pH of the solution is then equal to the negative logarithm of [H^+]. For example, a solution containing 0.01 mole of HCl in a liter of water (0.01 M $\cdot$ HCl) contains 0.01 M H^+ and 0.01 M Cl^-. The pH of this solution is 2 ($-\log_{10}$[0.01 M H^+]). Complete dissociation of a strong base yields a [$^-$OH] equal to the initial concentration of the strong base. The pH of the solution equals $14 - (-\log_{10}$[$^-$OH]). For example, a solution containing 0.001 mole of NaOH in a liter of water (0.001 M NaOH) contains 0.001 M $^-$OH and 0.001 M Na^+. The pH of this solution is 11 [$14 - (-\log_{10} 0.001$ M $^-$OH)].

Computation of the pH for a Solution of a Strong Acid

Since a strong acid is completely dissociated, the computation of the pH of a solution of a strong acid is simple and straightforward.

Example: Compute the concentration of the hydrogen ion in moles/liter (M) and then calculate the negative logarithm of that concentration; that is, if you add 0.4 mmoles of a strong acid to 100 mL of water, what is the pH?

$$100 \text{ mL} = 0.1 \text{ liters of water}$$

Therefore, the hydrogen ion concentration is 0.4 mmoles/0.1 liter or 0.4×10^{-3} moles/0.1 liter or 4×10^{-3} M

$$-\log_{10}(4 \times 10^{-3} \text{ M}) = 2.39$$

Thus the pH is 2.39.

---

[4] The counter ion is sometimes called the conjugate acid.

Weak Acids and Bases

In contrast to strong acids, weak acids do not completely dissociate when added to water. Thus the concentration of H^+ released from a weak acid is not equivalent to the initial concentration of the weak acid, since some protons are still bound in the undissociated weak acid. As an example, take acetic acid (CH_3COOH) (See Fig. 2.4). When added to water, only some of the acetic acid dissociates to produce H^+ and the conjugate base (acetate anion, CH_3COO^-); the remainder is undissociated acetic acid. The calculation of the pH of solutions of weak acids and their salts is best done in the discussion of pH buffers.

The strength of a weak acid is determined by the chemical structure of the acid. The degree of dissociation of an acid is represented by an equilibrium (dissociation) constant K_a, where the "a" defines this as an acid dissociation constant.

$$K_a = \frac{[H^+][\text{conjugate base}]}{[\text{acid}]}$$
$$K_a = \frac{[H^+][\text{acetate anion}]}{[\text{acetic acid}]} \tag{2.5}$$

The larger the value of K_a, the stronger the acid. Similar to pH notation, the acid dissociation constant, K_a, is usually expressed in terms of a pK_a where the pK_a is defined as the negative logarithm of the acid dissociation constant.

$$pK_a = -\log_{10}(K_a) \equiv \log\left(\frac{1}{K_a}\right) \tag{2.6}$$

For example, the K_a for acetic acid is 1.74×10^{-5}, and its pK_a is 4.76. As the K_a increases, the pK_a decreases; or said in another way, the lower the pK_a, the stronger the acid.

Some acids have multiple ionizable functional groups and thus have multiple pK_as. Such compounds are called polyprotic acids. To differentiate the different pK_a values, they are designated with numerical subscripts with the strongest acid group, lowest pK_a, designated as pK_1 and the weakest acid group, highest pK_a,

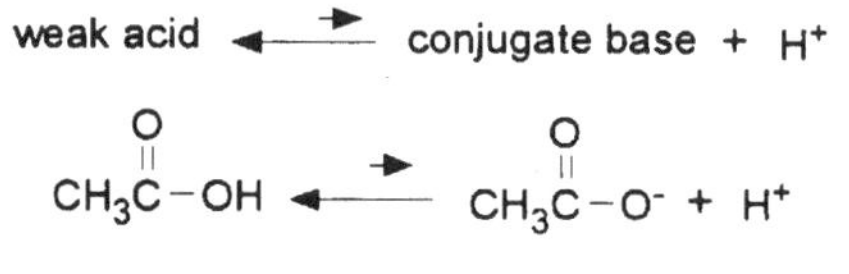

weak acid ⇌ conjugate base + H⁺

$$CH_3\overset{O}{\overset{\|}{C}}-OH \rightleftharpoons CH_3\overset{O}{\overset{\|}{C}}-O^- + H^+$$

Fig. 2.4. Dissociation of a Weak Acid.

Volt Meter

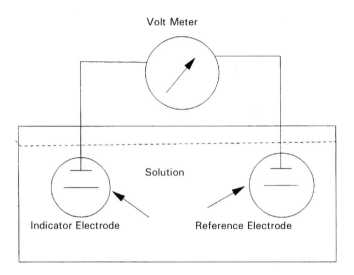

Fig. 2.5. A schematic of a pH meter measurement system.

having the highest numerical subscript. For example, phosphoric acid is a triprotic acid, has three dissociable protons, and has three pK_a values, pK_1, pK_2, and pK_3.

pH Measurements

Most pH measurements use pH meters[5] with a pH electrode (glass electrode) and a reference electrode. The Ag/AgCl electrode and the calomel electrode are the most common reference electrodes for pH measurements. A simplified diagram of a typical pH measurement system is shown in Figure 2.5.

The total voltage (potential) of such a pH measurement system (Eqn. 2.7) is the sum of the voltage from the reference electrode, the voltage from the indicator electrode, and the voltage that results from the interface of these electrodes with the solutions (junction potential).

$$E_{total} = E_{indicator} - E_{reference} + E_{junction} \tag{2.7}$$

Junction potentials are significant in size and may vary between electrodes and even with the same electrode depending upon the chemical characteristics of the sample solutions. Given the variability and lack of predictability of the junction potential, it is always necessary to calibrate a pH meter. Calibration is done with

---

[5] The use of pH indicators to measure pH will be discussed in Chapter 3.

solutions of known pH, usually near the pH of the desired measurements. Commercial solutions are available for calibrations at pHs of 4.0, 7.0, and 10.0. pH measurements are very difficult to do at low ionic strength (water and very low salt concentration <0.01 M) due to drifting of the pH meter. The addition of a neutral salt (e.g., $NaNO_3$) does not alter the $[H^+]$, brings the ionic strength to some common fixed level (typically 0.1 M), and prevents the drifting of the pH meter.

Acid–Base Indicators

A number of weak acids change color when they are protonated or undergo deprotonation. Such weak acids are called acid–base indicators or pH indicators. Acid–base indicators have multiple uses, such as determining the endpoints of acid–base titrations and determining the pH of water systems such as swimming pools and hot tubs. Table 2.2 has a list of some of the more common acid–base indicators and their respective useful pH ranges. Acid–base indicators are the active components of pH papers.

Titrations

While pH meters and pH papers are commonly used to determine hydrogen ion concentration, such techniques are somewhat lacking in precision. When really precise measurements are needed, then the analyst turns to acid–base titrations.

Definition: A titration is a quantitative analytical method wherein the concentration of the analyte solution is determined by adding sequential volumes of a reagent of known concentration to a solution of the analyte and determining that volume of reagent wherein there is a stoichiometric conversion of the analyte to a given product. The original analyte concentration is a function of the reagent

TABLE 2.2. Some Common Acid–Base Indicators and Their Properties

Indicator	Acid Color	Basic Color	pH Range
Cresol red (acid range)	red	yellow	0.2–1.8
Thymol blue (acid range)	red	yellow	1.2–2.8
Bromophenol blue	yellow	purple	3.0–4.6
Congo red	violet	reddish orange	3.0–5.0
Bromocresol green	yellow	blue	3.6–5.2
Methyl red	red	yellow	4.2–6.3
Cresol red (basic range)	yellow	red	7.2–8.8
Thymol blue (basic range)	yellow	blue	8.0–9.6
Phenolphthalein	colorless	purple	8.3–10.0

Source: Ref. 2.

concentration and the volume of reagent required to effect the stoichiometric conversion of the analyte to the product.

Titrations are attractive because they require only modest technical skill, are relatively fast assays, and require reasonably inexpensive equipment. Numerous different titrations have been described in the scientific literature for the analysis of a large number of analytes. Titrations are among the few assays wherein the analyst can directly view the stoichiometry of chemical reactions. Titrations are less useful when the transition of the observed signals are not abrupt, that is, when the assay reactions are relatively slow, or when the equilibrium constants lead to incomplete reaction. The appropriate design of blanks can be quite difficult with titration systems. The complexities of many biological systems are such that many titration assays cannot be readily used to determine the concentration of most analytes in biological systems.

Acid–Base Titrations

If one is adding acid to a base solution or base to an acid solution, the process is called an acid–base titration. Reactions of free H^+ and OH^- ions are extremely fast. Furthermore, the ion product of water is small ($K_w = [H^+][OH^-] = 10^{-14} \, M^2$). Thus as a first approximation, if a moderately concentrated solution of acid is added to a solution containing base, the free H^+ will react with free OH^- ions until either the free H^+ or free OH^- ions are consumed. If one continues to add H^+, then eventually all the OH^- ions are consumed and the pH will undergo an abrupt change from being basic to being acid. The change in pH in an acid–base titration is usually observed with a pH meter or with a pH indicator. A typical acid–base titration of a strong acid is shown in Figure 2.6. A typical titration of a weak acid is shown in Figure 2.7. The point in a titration where the amount of added $H^+(OH^-)$ equals the amount of base (acid) in the solution is called the *equivalence point*. In the titration shown in Figure 2.6 the equivalence point is at pH 7.0. Typical acid–base titrations of polyprotic acids are shown in Figures 2.8 and 2.9. Note that the change in pH with the addition of strong base is less in pH ranges near the pK_a values ($pK_1 = 2.12, pK_2 = 7.21, pK_3 = 12.3$) than in other areas. Phosphoric acid and its salts can be used to make buffers near each of these pK_a values.

Definition: The point in a titration where the molar amount of added reagent equals the molar amount of a particular analyte is called the equivalence point[6].

For a pH titration system to be successful as a quantifying tool, the signal being observed must undergo a rapid change at or very near the equivalence

---

[6] Assuming a one-to-one stoichiometry between the analyte and the titrant.

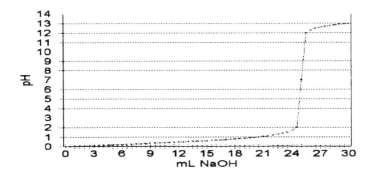

Fig. 2.6. Titration of a strong acid with a strong base.

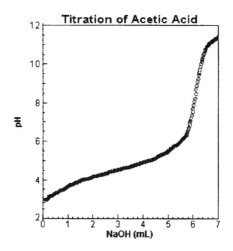

Fig. 2.7. Titration of a weak acid with a strong base.

point. Such is the case with titrations of strong acids with strong bases or visa versa. Thus titrations are often used to determine the concentrations of strong acids or bases. The titrations of weak acids or bases may, or may not, produce rapid changes in the signal at or near the equivalence points, and titrations are less frequently used to determine the hydrogen ion concentrations of solutions of weak acids. The titration curves of weak acids and mixtures of weak acids can be complex. See, for example, Figure 2.8, a titration curve of citric acid. Note that while citric acid is known to have three ionizable hydrogens, there are no breaks in the titration curve of citric acid. The lack of breaks occurs because the pK_as of citric acid are too close to each other.

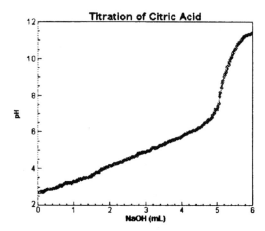

Fig. 2.8.   Titration of a polyprotic acid with NaOH.

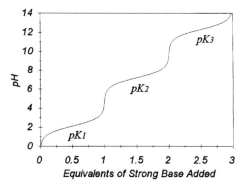

Fig. 2.9.   Titration of phosphoric acid (H_3PO_4) with strong base (e.g., NaOH). Note that the change in pH with strong base addition is less in pH ranges near the pK_a values ($pK_1 = 2.12$, $pK_2 = 7.21$, $pK_3 = 12.3$) than in other areas. Phosphoric acid and its salts can be used to make buffers near each of these pK_a values.

BUFFERS

pH Buffers

Many chemical and biochemical reactions produce or consume hydrogen ions. The production of hydrogen ions in these reactions is equivalent to the addition of a strong acid (e.g., HCl), and the consumption of hydrogen ions is equivalent to the addition of a strong base (e.g., NaOH). In vivo and in vitro systems frequently need to be protected from the reactivity of H^+ and ^-OH. This protection is usually accomplished using a pH buffer. A pH buffer is frequently used as a part of chemical measurements in biological systems. It is important for analysts to understand the factors that control the H^+ concentrations of buffer solutions, to know how to prepare buffers, and to understand the uses and limitations of pH buffer systems.

pH buffers are a mixture of one or more weak acids and their conjugate bases, and they are resistant to changes in pH upon the addition of protons or bases in a range of ± 1 pH unit of the pK_a of the weak acid used to prepare the buffer. Thus weak acids with different pK_as are used to prepare buffers of different pHs. When protons (H^+) are added to a buffer, they react with the conjugate base to form the weak acid; added ^-OH ions react with the weak acid to form the conjugate base.

In the following discussion, the weak acid component of the acid–base pair will be designated HA (the protonated form), and the weak base (conjugate base) component will be designated (A^-) (the unprotonated form). The protonated form, HA, dissociates to yield the unprotonated form (A^-) and a proton (H^+). Note that these designations are very general because they do not specify a charge of the protonated and unprotonated forms of the buffer system. The protonated form will always have a more positive charge than the unprotonated form. However, depending upon the specific buffer system selected, the protonated form may be positively charged (e.g., the ammonium ion), neutral (e.g., acetic acid), or negative (e.g., bicarbonate). The unprotonated form may have a net positive charge (e.g., the singly protonated form of ethylenediamine), a neutral charge (e.g., ammonia), or a negative charge (e.g., acetate anion).

In buffer solutions, the hydrogen ion concentration[7] ([H^+]) is dependent upon K_a (a constant for each weak acid) and the ratio of the concentration of weak acid and conjugate base.

$$[H^+] = K_a \frac{[HA]}{[A^-]} \qquad (2.8)$$

Note that this equation is a variation of Eq. 2.5 using the [HA] and [A^-] nomenclature to replace the weak acid and conjugate base, respectively. Addition of H^+ or ^-OH to a buffer solution changes the [A^-]/[HA] ratio and therefore the [H^+]. The buffer solution will thus resist changes in pH upon the addition of either strong acid or base. In contrast, when an equivalent amount of H^+ or ^-OH is added to water, the change in pH is much larger. Typical changes in pH when H^+ or ^-OH are added to water or a buffer solution are shown in Figure 2.10. It is important to recognize that adding strong acid or strong base to a solution will always result in a pH change. The presence of a buffer in the solution only limits the size of the resulting pH change.

---

[7] Note that the convention is to use [H^+] to designate the hydrogen ion concentration and to use (H^+) to just designate the species as a hydrogen ion.

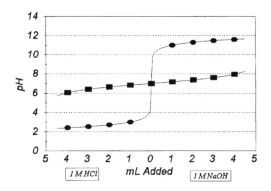

Fig. 2.10. Effect on pH of adding strong acid (1 M HCl) or strong base (1 M NaOH) to 1 liter of pure water (●) or 1 liter of 10 mM imidazole buffer (■) ($pK_a = 7.0$). The center position (0) on the x-axis is either pure water (●) or an equal molar mixture of the weak acid and conjugate base of the imidazole buffer (■). In either case, the pH of the solution prior to adding strong acid or strong base is 7.0.

Henderson–Hasselbach Equation

The relationship between pK_a, pH, [HA], and [A$^-$] for a buffer system is often called the Henderson–Hasselbach equation.

$$pH = pK_a + \log_{10} \frac{[A^-]}{[HA]} \tag{2.9}$$

The Henderson–Hasselbach equation is a logarithmic transformation of acid dissociation equation (Equation 2.8) in which pH is substituted for $-\log[H^+]$ and pK_a is substituted for $-\log K_a$. Using the Henderson–Hasselbach equation, the ratio of conjugate base to weak acid ([A$^-$]/[HA]) can be computed if the pK_a and pH are known. If the pK_a, [A$^-$] and [HA] are known then the pH can be computed. Thus, the Henderson–Hasselbach equation can be used to predict the effect variation in [A$^-$] and [HA] has on the pH of the solution or the effect the pH of the solution has on the variation in [A$^-$] and [HA].

Several special observations are worth noting.

1. If equal concentrations of a weak acid and its conjugate base are added to water (i.e., [HA] $=$ [A$^-$]), the pH of the solution should be equal to the pK_a of the weak acid because $\log_{10}(1) = 0$. For example, combining equal molar amounts of acetic acid and sodium acetate results in a solution with pH $= 4.76$, the pK_a of acetic acid.

2. If [A$^-$] $>$ [HA], then the ratio of [A$^-$] to [HA] is greater than 1, the $\log_{10}$ ([A$^-$]/[HA]) is positive, and the pH is greater than the pK_a of the weak acid.

3. If the $[A^-] < [HA]$, the value of $[A^-]/[HA]$ is less than 1, the $\log_{10}$ ($[A^-]/[HA]$) is negative, and the pH is less than the pK_a of the acid.

4. The Henderson–Hasselbach equation cannot be used effectively to compute the pH of a solution prepared by the addition of only a weak acid (or the conjugate base of a weak acid) to water. The solution is acid because the weak acid dissociates to a limited extent to produce protons (H^+). However, since only the pK_a is known, when one attempts to use the Henderson–Hasselbach equation, there are two unknowns (pH and $[A^-]/[HA]$) and only one equation. (Remember, one cannot solve for two unknowns using a single equation.) The estimation of the pH of a solution prepared by the addition of only a weak acid to water is complex and beyond the scope of this text.

5. Changes in $[A^-]$ and $[HA]$ upon the addition of strong acid (or strong base) are reciprocal; that is, increases in $[HA]$ are equal to decreases in $[A^-]$ and vice versa. It is a "zero sum game"; the total concentration of the buffer ($[A^-] + [HA]$) is constant; only the distribution between the unprotonated and protonated forms changes.

6. Dilution of a buffer solution with pure water should not change its pH. While dilution decreases $[A^-]$ and $[HA]$, their ratio is not altered. Since it is the ratio $[A^-]/[HA]$ that is important in the Henderson–Hasselbach equation, the pH of the solution should not be affected by dilution.[8]

7. It is normally assumed that the useful range of a buffer solution varies from the pH at which $[A^-]/[HA]$ is $1/10$, and the pH at which $[A^-]/[HA]$ is $10/1$. If the ratio is less than $1/10$ or greater than $10/1$, the buffer solution has limited capacity to resist pH changes upon the addition of H^+ or ^-OH. Using these ratios and the Henderson–Hasselbach equation, the useful range of a buffer is the $pK_a \pm 1$ pH unit.[9] For example, acetic acid has a pK_a of 4.76; so the useful pH range for an acetate buffer would be 3.76–5.76.

8. The pK_a values of several weak acids are temperature dependent. Thus the final pH of a buffer should be checked at the temperature at which the buffer will be used.

Buffer Selection

The selection of the appropriate pH for chemical measurements is dependent upon several factors. With in vitro enzyme activity measurements, for example, these factors include the stability of the enzyme as a function of pH, the influence of pH on enzymatic activity, and the ionization state(s) of the substrate(s) and product(s) of the enzyme-catalyzed reaction. For most chemical analyses in

[8] There are often small changes in pH upon dilution because of the presence of H_2CO_3 in the water.

[9] Some analysts state that a useful range of a buffer is $pK_a \pm 2$ pH units, but the capacity to resist changes in pH is much poorer when the ratio is between $1/100$ and $1/10$ or $100/1$ to $10/1$.

biological systems the selection of an appropriate buffer is empirical. However, there are several general considerations. The buffer system should have a pK_a close to the desired pH (i.e., within one pH unit of the pK_a), the buffer components should not interfere with the reaction chemistry or the detection process,[10] and the buffer pH should be measured at the temperature of the reaction, since some buffers have temperature-sensitive pK_a values. Buffer concentration is often important. While the concentration of the buffer should be sufficient to resist pH changes in the system, "the higher the better" may not be appropriate in the analyses of biological systems. The catalytic activities of some enzymes are significantly influenced by the ionic strength of the buffer solution. Buffer concentrations between 10 and 100 mM are the most frequently used. If intact cells are present in the assay mixtures, osmotic pressure effects need to be considered.

There are a variety of literature sources that provide lists of buffers and pK_a values.[11] Such references can be useful as recipe books for buffer preparation. However, it should be noted that sometimes the pK_a values presented in these lists are the thermodynamic pK_a values. These thermodynamic pK_a values are established at infinite dilution of the buffer, where the ionic activities and concentration are identical. At buffer concentrations useful for biological systems, the "functional" pK_a may be different from the thermodynamic pK_a. For example, the thermodynamic pK_2 of phosphoric acid is 7.2, but the "functional" pK_2 is 6.8. The thermodynamic pK_a values of buffer systems commonly used in chemical measurements in biological systems are given in Table 2.3. Polyprotic acids (e.g., citric acid) can serve as buffers in pH ranges near each of their pK_a values. Titration curves of polyprotic acids may be more complex than those of monoprotic acids (Figs. 2.8 and 2.9).

Buffer Preparation

Buffer preparation is a recurrent task in laboratories doing chemical measurements in biological systems. Four pieces of information are needed to make a buffer: (1) the final buffer volume, (2) the final pH, (3) the total molar concentration of the buffer,[12] and (4) the weak acid–conjugate base combination to be used in the buffer.

A buffer solution could be prepared by calculating the appropriate amounts of acid and conjugate base components of a buffer based on the desired pH, the

---

[10] Common problems include chelation of metals by some buffer component, inhibition of enzymatic reactions, production of gases or precipitates, and other interferences with the detection process.

[11] See, for example, Ref. 2.

[12] Remember that the total concentration of a buffer = [weak acid] + [conjugate base].

TABLE 2.3. pK_a Values of Some Common Buffer Systems

Buffer Name	pK_1	pK_2	pK_3
ACES [N-(2-acetyamido)-2-aminoethanesulfonic acid]	6.9		
Acetic acid	4.76		
Barbital (5,5-deithylbarbituric acid)	7.78		
Bicine [N,N-bis(2-hydroxyethyl)glycine]	8.35		
CAPS [3-(cyclohexylamino)-1-propanesulfonic acid]	10.4		
Carbonic acid	6.35	10.33	
CHES [2-(N-cyclohexylamino)ethanesulfonic acid]	9.3		
Citric acid	3.13	4.76	6.4
Ethanolamine	9.5		
Ethylenediamine	6.85	9.93	
Formic acid	3.75		
Glycine	2.35	9.78	
Glycylglycine	3.14	8.25	
HEPES (N-2-hydroxyethylpiperazine-N'-2-ethanesulfonic acid)	7.55		
MES [2-(N-morpholino)ethanesulfonic acid]	6.15		
MOPS (4-morpholinopropanesulfonic acid)	7.2		
Phosphoric acid	2.12	7.21	12.3
PIPES			
[piperazine-N,N'-bis(2-ethanesulfonic acid)]	6.8		
Succinic acid	4.19	5.57	
TES			
[N-tris(hydroxymethyl)methyl-2-aminoethanesulfonic acid]	7.5		
Tricine			
[N-tris(hydroxymethyl)methylglycine]	8.15		
Triethanolamine	7.8		
Triethylamine	10.7		
TRIS [tris(hydroxymethyl)aminomethane]	8.3		

literature value of pK_a the desired volume, and the selected buffer concentration. However, most analysts do not prepare buffers in this manner.

Method One—Titration

In this method, a solution of the weak acid (e.g., acetic acid) is prepared. The concentration of this solution is slightly higher than that in the final buffer. This weak acid solution is titrated with a strong base (e.g., NaOH) to the desired pH using a pH meter. The titrated solution is diluted to the final volume and concentration, and the final pH of the buffer is measured and adjusted, if necessary. Similarly, a weak base (e.g., Tris) can be titrated to a desired pH with a strong acid (e.g., HCl). Remember that buffer pHs should always be checked at the temperature where they will be used.

Method Two—Use Literature "Recipe" for Buffer Preparation

The preparation of a buffer can be accomplished by combining solutions of the weak acid and the salt of the conjugate base of the weak acid. For example, an acetate buffer may be prepared by adding a solution of sodium acetate to a solution of acetic acid. As mentioned earlier, there are several literature sources that provide recipes for the preparation of buffers. Often such references have the analyst prepare solutions of the weak acid and the conjugate base at the final concentration of the buffer. These solutions are then mixed in the proper proportions to prepare the buffer. To the first approximation, the salt of the conjugate base is completely dissociated in solution whereas the weak acid is assumed to be undissociated. Using these assumptions, the Henderson–Hasselbach equation can be rewritten as:

$$pH = pK_a + \log_{10} \frac{Vol_{A^-}}{Vol_{HA}} \qquad (2.10)$$

A few simple calculations will show that when each part (weak acid and conjugate base) of the buffer is prepared at the final concentration, then the added volumes of these solutions may be used to calculate the final pH of the buffer using the Henderson–Hasselbach equation.

Method Three—Calculate the Required Volumes

Calculation of the volumes of the weak acid and its conjugate base needed to prepare a buffer is perhaps the most difficult way to prepare a buffer. As with Method Two, the analyst prepares solutions of the weak acid and the conjugate base, each at the final concentration of the buffer. This approach makes the computations relatively simple. The following is an example of this method.

Assume that you need 10 mL (0.01 L) of 0.2 M acetate buffer, pH 5.0. Prepare solutions of 0.2 M acetic acid (weak acid) and 0.2 M sodium acetate (conjugate base). The pK_a of acetic acid is 4.76 (Table 2.4). Knowing the pH of the desired buffer solution and the pK_a of the weak acid allows the calculation of the ratio of conjugate base [A$^-$] to weak acid (HA) from the Henderson–Hasselbach equation (Eq. 2.9).

$$pH = pK_a + \log \frac{[A^-]}{[HA]}$$

$$5.0 = 4.76 + \log_{10} \frac{[A^-]}{[HA]}$$

$$\log_{10} \frac{[A^-]}{[HA]} = 5.0 - 4.76 = 0.24$$

$$\frac{[A^-]}{[HA]} = \text{anti log } 0.24 = 1.74$$

$$[A^-] = 1.74[HA]$$

Thus the amount of the conjugate base ($[A^-]$, sodium acetate in this example) must be 1.74 times that of the weak acid ($[HA]$, acetic acid in this example). Since the solutions of acetic acid and sodium acetate were of equal concentration, the volume of sodium acetate required to prepare the desired buffer must be 1.74 times the volume of acetic acid. Remember that the final volume of the buffer needed is 10 mL.

$$\text{Volume}_{A^-} \text{(sodium acetate)} = 1.74 \text{ Volume}_{HA} \text{ (acetic acid)}$$

$$\text{Total volume} = 10 \text{ mL} = \text{Volume}_{A^-} + \text{Volume}_{HA}$$

$$= 1.74 \text{ Volume}_{HA} + \text{Volume}_{HA} = 2.74 \text{ Volume}_{HA}$$

$$\text{Volume}_{HA} = 10 \text{ mL}/2.74 = 3.65 \text{ mL}$$

$$\text{Volume}_{A^-} = 10 - 3.65 \text{ mL} = 6.35 \text{ mL}$$

As with the preparation of buffer using any method, the final step should be to check the pH of the buffer at the temperature where it will be used and make any necessary adjustments.

Nomenclature

The protocol for naming buffer solutions may be confusing. However, it is important that the analyst understand it. For example, a 0.1 M buffer solution made by titrating Tris,[13] a weak base, to pH 8.0 with HCl, a strong acid, is designated as 0.1 M Tris-Cl, pH 8.0. This label means that the total buffer concentration, Tris (the weak base) + Tris-HCl (the conjugate acid), is 0.1 M, that the solution has a pH of 8.0, and that the counter ion is Cl^-. It is *not* a solution of 0.1 M Tris·hydrochloride *nor* is the Cl^- concentration specified. The counter anion may also be shown in parentheses; for example, 0.1 M Tris(SO_4), pH 8.0, represents a 0.1 M solution of Tris at pH 8.0 with sulfate as a counter ion. Similarly, a buffer made from a weak acid, such as phosphoric acid, may be designated 0.05 M KPO_4, pH 7.0. This label means that the total concentration of phosphate $(H_3PO_4 + H_2PO_4^- + HPO_4^{2-} + PO_4^{3-})$[14] is 0.05 M and that the counter ion is the potassium ion. The concentration of the counter ion (K^+) is *not* designated. The ionic strength of these buffers is not designated, nor is it easy to compute. When higher-ionic-strength buffer is needed (e.g., in liquid chromatography), the usual practice is to add neutral salts such as KCl or NaCl to a given concentration. A buffer labeled 0.1 M Tris(Cl), pH 8.0, 0.15 M NaCl signifies that

---

[13] "Tris" is an abbreviation for tris(hydroxymethyl)aminomethane, which is a common buffer component.

[14] The specific concentration of each of these ionic species of phosphate is a function of the pH.

the total concentration of Tris buffer (weak base + conjugate acid) is 0.1 M, that
the solution has a pH of 8.0, and that the concentration of the counter ion Cl^- is
0.15 M (from the NaCl) *plus* the concentration of the Cl^- present in the Tris(Cl)
buffer.

Buffer Capacity

The ability of a buffer to resist a change in pH upon the addition of strong acid
(H^+) or strong base (^-OH) is called its *buffer capacity*. In a buffered solution, the
size of the pH change resulting from the addition of H^+ or ^-OH is dependent
upon the initial $[A^-]/[HA]$ ratio (a function of the pH and pK_a), the buffer
concentration, the amount of H^+ or ^-OH added, and whether H^+ or ^-OH is
added.[15] Buffer capacity is formally defined as the amount of strong acid or
strong base that must be added to a buffer to decrease or increase the pH by one
unit, respectively. The greater the amount of acid or base required to change the
pH by one unit, the greater the buffer capacity of the system.

Usually, the analyst is more interested in whether the buffer will maintain pH
within given limits and less interested in the actual value of buffer capacity for a
specific system. Determination of whether the buffering capacity of the system
has been exceeded is relatively straightforward. The following steps can be used
to decide if a buffer would still have some buffering capacity after the addition of
some hydrogen ions (H^+).

1. Estimate the *amount* of added H^+ as follows:

$$H^+ \text{amount} = [\text{concentration of added } H^+] \times [\text{volume of added } H^+]$$

2. Estimate the *amount* of conjugate base in the volume of buffer to which the
H^+ was added. This amount of conjugate base in the buffer is computed as
follows:[16]

$$\text{Amount of conjugate base } [A^-] \text{ in buffer} =$$
$$[\text{concentration of } A^- \text{ in buffer}] \times [\text{volume of buffer}]$$

If the *amount* of added H^+ is greater than the *amount* of conjugate base in the
buffer, then the buffer capacity of the system has been exceeded. If the *amount* of

---

[15] The buffering capacity for H^+ or ^-OH addition will be the same only when $[A^-] = [HA]$, i.e.,
when pH = pK_a.

[16] The concentration of the conjugate base $[A^-]$ is dependent upon the pH of the solution and the
pK_a of the weak acid.

added H^+ is less than the *amount* of conjugate base in the buffer, then the buffer capacity has not been exceeded.

Since the buffer capacity is directly dependent upon the concentration of the conjugate base, the buffer capacity must be dependent upon the buffer concentration. Thus any dilution of the buffer will reduce the buffer capacity. For example, a 0.01 M Tris(Cl), pH 8.0, buffer would have 1/10 the buffer capacity of 0.1 M Tris(Cl), pH 8.0, buffer.

EXAMPLES OF pH AND BUFFER PROBLEMS

Example 1. The change in pH with the addition of H^+ to water.
Background information:

$$\text{Volume of water} = 100 \text{ mL}$$

$$\text{Volume of strong acid added} = 0.1 \text{ mL}$$

$$\text{Concentration of strong acid} = 0.1 \text{ M}$$

$$[H^+] = \frac{(\text{volume}_{acid})([\text{acid}])}{\text{volume}_{H_2O} + \text{volume}_{acid}}$$

$$[H^+] = \frac{(0.1\text{mL})(0.1\text{M})}{100\text{mL} + 0.1\text{mL}} = 9.99 \times 10^{-5}\text{M}$$

$$pH = -\log_{10}[H^+] = \log_{10}(9.99 \times 10^{-5}\text{M}) = 4.0$$

Example 2. The change in pH with the addition of strong acid (H^+) to a buffer.

a. When the amount of added H^+ is less than the amount of conjugate base $[A^-]$ in the original buffer. The information needed to make this calculation includes the initial buffer pH, the buffer concentration, the buffer volume, the pK_a of the weak acid, and the concentration and volume of the strong acid added. It is assumed that all the H^+ from the strong acid combines with the conjugate base $[A^-]$ of the buffer to form the weak acid of the buffer (HA). Thus, for each mole of H^+ added, the amount of conjugate base A^- decreases by a mole and the amount of weak acid HA increases by a mole.

Background information:

$$\text{Volume of buffer} = 100 \text{ mL}$$

$$\text{Concentration of buffer} = 0.2 \text{ M}$$

$$\text{Initial pH} = 7.5$$

$$\text{p}K_a \text{ of weak acid} = 7.0$$

$$\text{Concentration of strong acid} = 2\text{M}$$

$$\text{Volume of strong acid added} = 5 \text{ mL}$$

i. Calculation of moles of conjugate base (A^-) and weak acid (HA) in the buffer before the addition of the strong acid.

The ratio of conjugate base to weak acid concentration ($[[A^-]]/[HA]$) in the buffer before adding strong acid is calculated using the Henderson–Hasselbach equation (Eq. 2.9).

$$[A^-]/[HA] = 10^{(\text{pH}-\text{p}K_a)}$$

$$[A^-]/[HA] = 10^{(7.5-7.0)} = 10^{(0.5)} = 3.16$$

$$[A^-] = 3.16[HA]$$

The moles of buffer $(A^- + HA)$ are found from the volume of the buffer and its concentration.

$$\text{moles buffer} = [\text{buffer concentration (M)}] \times [\text{buffer volume (L)}]$$

$$= (0.2 \text{ M}) \times (0.1 \text{ L}) = 0.02 \text{ moles}$$

Since moles of $[A^-] + [HA] = 0.02$ moles and $[A^-] = 3.16 \times [HA]$, $[HA] = 0.02$ moles/4.16 $= 0.0048$ moles and $[A^-] = 0.0152$ moles before adding the strong acid.

ii. Calculation of moles of strong acid added.

$$\text{moles acid} = [\text{acid concentration (M)}] \times [\text{acid volume(L)}]$$

$$\text{moles acid} = (2.0 \text{ M}) \times (0.005 \text{ L}) = 0.01 \text{ moles}$$

iii. Calculation of moles of conjugate base $[A^-]$ and weak acid (HA) in the buffer after adding strong acid.

It is assumed that strong acid combines with the conjugate base to produce the weak acid on a one-for-one basis. Thus, after the addition of strong acid, there are 0.0052 moles of conjugate base (0.0152 moles− 0.01 moles) and 0.0148 moles of weak acid (0.0048 moles + 0.01 moles).[17]

iv. Calculation of pH after the addition of strong acid.

The pH of the solution after the addition of strong acid is calculated using the Henderson–Hasselbach equation (Eq. 2.9). The pK_a is constant (7.0), but $[A^-]/[HA]$ changes.

$$pH = 7.0 + \log_{10} \frac{0.0052 \text{ moles}}{0.0148 \text{ moles}} = 7.0 + \log_{10}(0.35) = 7.0 - 0.45 = 6.55$$

Thus adding 0.01 moles of strong acid to this buffer solution causes the pH to decrease from 7.5 to 6.55.

b. When the amount of added H^+ is greater than the amount of conjugate base $[A^-]$ in the original buffer. The information needed to make this calculation is the same as that needed in *a*.

Background information:

Volume of buffer = 100 mL

Concentration of buffer = 0.2 M

Initial pH = 7.5

pK_a of weak acid = 7.0

Concentration of strong acid = 2 M

Volume of strong acid added = 10 mL

The initial steps involved in the calculation are comparable to those in a. In this example, the amount of strong acid added is 0.02 moles (2 M × 0.01 L), which exceeds the amount of conjugate base (0.0152 moles) in the buffer solution before adding acid (see a1). The 0.0152 moles of conjugate base can combine with 0.0152 moles of strong acid (H^+), but there is an excess of strong acid (0.048 moles). This excess of strong acid determines the pH of the solution.

$$pH = -\log_{10}[H^+] = -\log_{10} \frac{0.0048 \text{ moles}}{0.11 \text{ L}} = -\log_{10}(0.0436 \text{ M}) = 1.36$$

[17] Note that the total amount of buffer (0.02 moles) remains the same. The amounts of weak acid and conjugate base change in opposite directions with the addition of strong acid.

REFERENCES

1. T. D. McCune, K. W. Lang, and M. P. Steinberg, *J. Food Sci.* **46**, 1978–79 (1981).
2. R. M. C. Dawson and D. C. Elliott, *Data for Biochemical Research*, 3rd Edition, Oxford University Press, Oxford, England, 1986.

3. Ultraviolet and Visible Absorption Spectrophotometry and Photometry

BACKGROUND

Spectrophotometric analysis (the study of the interaction of electromagnetic radiation with chemical compounds) is one of the most commonly used analytical techniques in the life sciences. Spectrophotometry can be used to determine the structure of a new compound, identify a specific compound, determine the concentration and/or amount of a particular compound (e.g., DNA, protein), and determine the activity of a specific enzyme. The electromagnetic spectrum ranges from cosmic rays with wavelengths of 10^{-6} nm (nanometers[1]), to radio waves with wavelengths of 10^{12} nm. Biological molecules interact in some way with many different parts of this spectrum.

Electromagnetic radiation behaves both as electromagnetic waves and as particles (photons). The distance between two adjacent wave peaks is defined as

[1] 10^{-9} meters.

the wavelength (λ) of the light. Wavelength has units of *meters*. The frequency (v)[2] defines the number of these waves that pass a particular point per unit time and has units of *cycles per second* (hertz, Hz). Wavelength and frequency are reciprocally related:

$$\lambda = \frac{c}{v} \qquad (3.1)$$

where λ is the wavelength in meters, c is the speed of light (3×10^{10} cm/sec), and v is the frequency of the light in cycles per second (hertz, Hz).

The energy of the photon is inversely proportional to the wavelength of the light:

$$E = h\frac{c}{\lambda} \qquad (3.2)$$

where E is the energy of the photon in joules, and h is Planck's constant in joules per second.

Thus the photons with very short wavelengths (X rays; 10^{-1} to 10^{-3} nm) have the highest energies and the photons with very long wavelengths (radio waves; 10^{0} to 10^{3} m) have the lowest energies. The photons of the visible and ultraviolet regions have intermediate energies.

The uv/vis region of the electromagnetic spectrum is the most commonly used region for quantitative chemical measurements in biological systems[3]. The ultraviolet (uv) region is defined as that region between 200 and 350 nm. The visible (vis) region (these wavelengths are visible to the human eye) is defined as that region between 350 to 750 nm. Light with wavelengths of 350–450 nm is blue, that of 450–550 nm is green, that of wavelengths of 550–650 nm is yellow, and that of 650–750 is red. Blue light has photons of the highest energy in the visible region, and red light has photons of the lowest energy in the visible region. The color of solutions and of reflected light is determined by the wavelengths of the light that are *not* absorbed by the solution or the compound reflecting the light. For instance, red food coloring appears red because the shorter wavelengths of light are absorbed by the colored solution. Green plants are green because

---

[2] In some applications, e.g., infrared spectroscopy, the wave number of light ($\bar{v}$) is used. Wave number is defined as the number of waves per distance (typically cm) and is the reciprocal of the wavelength.

[3] In addition to the uv-vis region, other areas of the electromagnetic spectrum are used to analyze compounds of biological interest. X rays are used to determine the precise location of atoms within a structure (X-ray crystallography). Infrared spectroscopy (IR), the absorption of the radiation of light with wavelengths between the visible and the microwave, is used to identify organic structures through their characteristic molecular vibrations and rotations. Microwave radiation is used in techniques that investigate nuclear and electronic spin (nuclear magnetic resonance, NMR, and electron spin resonance, ESR).

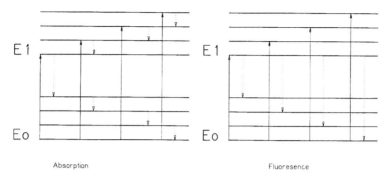

Fig. 3.1. Effect of Absorption of a Photon on the Energy Levels of a Compound. In absorption spectrophotometry, the absorption of a photon with the energy equal to the energy differences between the ground state (E_0) and an excited state (E_1) will excite an electron to that excited state. The electron will then lose energy through non-radiative process (e.g. thermal motion) and fall back to its original ground level energy. In fluorescent spectrophotometry, the absorption of a photon occurs by the same mechanism, however the electron will then lose energy through the emission of a photon and fall back to its original ground level energy.

chlorophyll absorbs light in the nongreen regions. However, for quantifying the concentration of an analyte, we are interested in the light that is *absorbed* by solutions, not in the light that is transmitted or reflected by the solution.

The process of absorption of light by a particular molecule is a discrete process. According to the quantum theory, every atom and molecule has a unique set of energy states, some filled and some empty (see Fig. 3.1). In the uv/vis region the absorption of light occurs when the absorbed photons excite an electron from one energy state to a higher-energy state. Once a photon has excited an electron to a higher-energy state, then that electron will lose energy through nonradiative relaxation due to molecular motion and collisions with other molecules, and eventually return to its original energy state. Alternatively this energy can be lost through emission of a lower-energy photon in a process called *fluorescence*.[4] The plot of absorption of the light by a compound versus the wavelength of the light is called the compound's *absorption spectrum*. A wavelength of maximum light absorption is called the λ_{max}. Many compounds have more than one λ_{max}.

The energy difference between the energy states of a given molecule is dependent upon the electronic structure of the specific compound. Since different molecules have different electronic states, the absorption of light by a molecule will be a function of both the electronic structure of the molecule and the wavelength of the light. Molecules with significant differences in their electronic structure will have markedly different absorption spectra; molecules with few differences in their electronic structure will have similar absorption spectra.

---

[4] The use of fluorescence as an analytical tool will be briefly discussed at the end of this chapter.

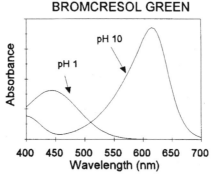

Fig. 3.2.   Structures of bromcresol green at pH1 and pH10.

Those changes in chemistries that alter the electronic structure of a molecule will alter its absorption spectrum. For example, the protonation of bromcresol green changes its electronic structure (see Fig. 3.2) and thus leads to a change in the spectra (see Fig. 3.3).

The electrons that are excited by light in the uv/vis region of the electromagnetic spectrum are those whose transitions require moderately low energy, such as those in outer shells of the transition metals and in the unsaturated bonds of organic compounds, particularly in conjugated systems. In the conjugated[5] organic compounds, the actual electronic transitions are either $\pi \rightarrow \pi^*$ or $n \rightarrow \pi^*$, where π represents π electrons in double bonds and n represents nonbonding electrons such as those in nonbonding electron pairs associated with oxygen in a carbonyl group. As the extent of conjugation increases, the energy difference between the ground and excited states is diminished and the energy of light needed to excite an electron from one state to the other decreases. Thus, as the conjugation of the system becomes more

BROMCRESOL GREEN

pH 10

pH 1

Absorbance

400   450   500   550   600   650   700
Wavelength (nm)

Fig. 3.3.   The spectra of the protonated (pH1) and unprotonated (pH10) forms of bromocresol green.

[5] A conjugated compound is one that has one or more sequences of a double bond–single bond–double bond. The electrons of the double bonds in these systems do not exist solely in the double bond but are delocalized over the entire double bond–single bond–double bond sequence.

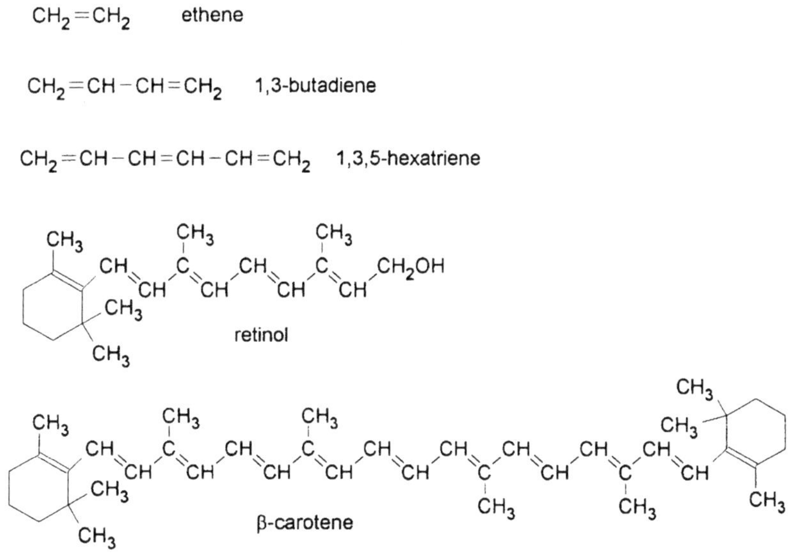

CH$_2$=CH$_2$ ethene

CH$_2$=CH–CH=CH$_2$ 1,3-butadiene

CH$_2$=CH–CH=CH–CH=CH$_2$ 1,3,5-hexatriene

retinol

β-carotene

Fig. 3.4. Some conjugated systems.

extensive, the wavelength of light absorbed by the system moves to longer wavelengths (lower energy). For example, compare the structures of ethylene, 1,3-butadiene, 1,3,5-hexatriene, retinol (vitamin A), and β-carotene (Fig. 3.4). As the number of double bonds in the conjugated system increases from one (ethylene) to eleven (β-carotene), the absorbance peak with the longest wavelength increases from 174 to 497 nm (Table 3.1). Retinol appears as a pale yellow color to the human eye, while β-carotene is the familiar orange color of carrots. In general, any structural change that increases the possibility of electron delocalization (resonance) decreases the energy needed to make the electronic transition and shifts the absorption of light to longer wavelengths (red or bathochromic shift).

For any individual molecule, the energy required to excite an electron from the ground to an excited state is a discrete value; however, actual analytical samples contain large numbers of these molecules.[6] If all the molecules or atoms have essentially identical energy levels, their observed spectra will be a line spectrum as is seen in the ionization of some gases and in atomic absorption spectroscopy (see Fig. 3.5). However, when these molecules are dissolved in solvents (such as water), which permits a large number of different orientations relative to the incident light and many possible vibrational modes at each electronic energy level, then each of these differences has an effect on the energy required to excite the electron so that the quantized energy differences will be spread out. Thus the

[6] For example, a one micromolar (μ)M solution has about 6×10^{14} molecules in a milliliter (mL).

TABLE 3.1.   Absorption of Conjugated Compounds

Compound	Number of Double Bonds in Conjugated System	Peak of Longest Wavelength Absorbed (nm)
Ethene[a]	1	174
1,3-Butadiene[b]	2	217
1,3,5-Hexatriene[c]	3	266
Retinol[b]	5	325
β-Carotene[d]	11	497

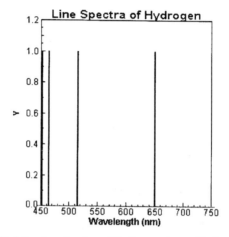

Fig. 3.5.   An estimation of the line spectrum of hydrogen.

spectrum of a solution of these molecules will be a smooth curve of multiple points rather than single lines (see Fig. 3.6).

In absorption spectrophotometry the intensity of the light at a given wavelength is determined at that wavelength at the detector when a sample is present in the light path and compared with the intensity of the light determined at the detector when there is no sample in the light path. The ratio of these two intensities is called the *transmittance*[7]

$$T = \frac{I_{\text{fnl}}}{I_{\text{o}}} \tag{3.3}$$

---

[7] A the key feature of Eq. 3.3 is that the change in the transmittance is not a function of the initial intensity of the light, and thus light absorbance measurements can be made with a variety of light intensities. These concepts were originally elucidated by Lambert and are the basis for Lambert's Law.

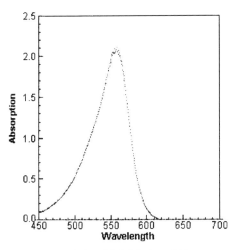

Fig. 3.6. Spectrum of phenolphthalein in a ph11 aqueous solution.

where T is the transmittance, I_{Fnl} the final intensity of light, and I_0 the initial intensity of light.

The negative logarithm of the transmittance of a solution is defined as the *absorbance*[8] A of a solution:

$$A = -\log T \tag{3.4}$$

The absorbance of light at any given wavelength is directly proportional to the number of absorbing molecules in the light path. Since the number of molecules in the light path is a direct function of the concentration of the analyte and the length of the light path, the absorbance of light by a solution is directly proportional to these variables. *Beer's Law*[9] states that the absorbance of light is proportional to the concentration of the absorbing species and the path length of the absorbing solution (l) by a constant called the *absorptivity* (a) if the concentration is expressed as g/L or the molar *absorptivity* (ε) if the concentration is expressed as moles/L.

$$A = alc = \varepsilon lc \tag{3.5}$$

The molar absorptivity is a function of the electronic structure of the molecule and the wavelength of the light passing through the solution. Absorptivity and

---

[8] Absorbance replaces the terms *optical density* and *extinction,* whose uses are no longer recommended.

[9] Beer's law has also been called the Lambert–Beer law.

molar absorptivity have units of reciprocal concentration (e.g., g/L^{-1} or M^{-1}) and reciprocal length (e.g., cm^{-1}). Note absorbance is a term without units.[10]

Most assays in biological systems are used to determine analyte concentrations in the milli- to nanomolar range. In such cases the use of the units of molar absorptivity can be rather awkward. One would much rather use molar absorptivity units closer to the mM or μM or nM concentrations of the compounds that are found in biological systems. Such conversions are straightforward, but can be slightly confusing to the novice. The "trick" to making this conversion is to recognize that *molar absorptivity units represent an absorbance per concentration*. Thus a molar absorptivity with units of $M^{-1}cm^{-1}$ represents the absorbance of a 1 M solution and a 1 cm light path length. A 1 mM solution of this compound should have an absorbance 1/1000 that of the 1 M solution; a 1 μM solution should have an absorbance 1/1,000,000 (10^{-6}) that of the 1 M solution, and a 1 nM solution should have an absorbance 1/1,000,000,000 (10^{-9}) that of the 1 M solution. For example, the literature value for the molar absorptivity at 340 nm of reduced nicotinamide adenine dinucleotide (NADH) is $6.22 \times 10^3 \ M^{-1}cm^{-1}$. The millimolar absorptivity of NADH is $6.22 \ mM^{-1} \ cm^{-1}$, while the micromolar absorptivity is $6.22 \times 10^{-3} \ \mu M^{-1} \ cm^{-1}$, and the nanomolar absorptivity is $6.22 \times 10^{-6} \ \mu M^{-1} \ cm^{-1}$. Sample unit conversions are shown:

$$6.22 \times 10^3 \ M^{-1} cm^{-1} = \frac{6.22 \times 10^3}{M \ cm} \frac{M}{10^3 \ mM} = \frac{6.22}{mM \ km} = 6.22 \ mM^{-1} \ cm^{-1}$$

(3.6a)

$$6.22 \times 10^3 \ M^{-1} cm^{-1} = \frac{6.22 \times 10^3}{M \ cm} \frac{M}{10^6 \ \mu M} = \frac{6.22}{\mu M \ cm} = 6.22 \ 10^{-3} \mu M^{-1} cm^{-1}$$

(3.6b)

CONCENTRATION CALCULATIONS

If the molar absorptivity at a specific wavelength is known for a specific compound, then calculation of the concentration of the compound can be determined from its absorbance at this wavelength using Beer's law. The concentration of the compound is equal to the absorbance divided by the molar absorptivity times the light path length:

$$c = \frac{A}{\varepsilon l}$$

(3.7)

---

[10] However, some chose to give it the units of AU (absorbance units) to make the unit analysis process easier.

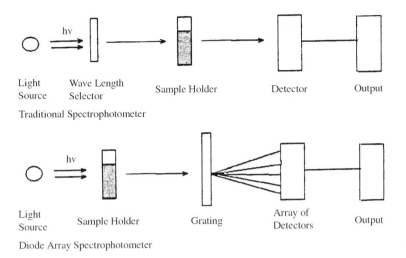

Light Source

Wave Length Selector

Sample Holder

Detector

Output

Traditional Spectrophotometer

Light Source

Sample Holder

Grating

Array of Detectors

Output

Diode Array Spectrophotometer

Fig. 3.7. Core components of spectrophotometers.

In those cases where the molar absorptivity is not known, it can be determined by determining the response curve for a spectrophotometric assay and then finding the straight-line portion of that response curve. The slope of the linear response curve is the absorptivity or molar absorptivity depending upon the units of the x axis. Like all response curves, the best way to find the linear portion of the response curve is to use the concepts inherent in the use of regression analysis with linear systems. Appendix 3 has a detailed discussion on the use of regression analysis.

The basic instrumentation for measuring the absorption of light is shown in Figure 3.7. While spectrophotometers have a variety of designs, they all have several components in common: a light source, some means of selecting the wavelength of the light, a sample cell, and a detector with some associated electronics. The common types of spectrophotometers are single-beam, double-beam, and diode-array spectrophotometers. With the classical single-beam and double-beam spectrophotometers, the wavelength of the light is selected before the light goes through the sample and one or two detectors are used. With a photodiode array spectrophotometer the wavelengths are selected after the light has passed through the sample and multiple detectors are used. Beer's law is equally applicable with all types of spectrophotometers.

The recent use of microtiter plates[11] for ELISA[12] assays, combinatorial chemistry, and other assays have resulted in the introduction of spectro-

---

[11] A microtiter plate is a 96-well reaction plate that was originally designed for immunological analyses (see Chapter 9).

[12] Enzyme-linked immunosorbant assay. See Chapter 9.

photometers for microtiter plates. These spectrophotometers are almost always single-beam instruments wherein the light is passed through the sample chamber from top to bottom (or reverse). These microtiter plate spectrophotometers are out of the ordinary in that the path length is usually not fixed and is a function of the volume of the sample in the microtiter plate well.

There are many types of light sources for spectrophotometric systems. For example, light emitting diodes (LEDs), lasers, deuterium lamps, carbon are lamps, tungsten–halogen lamps, tungsten lamps, hollow cathode lamps, and Hg lamps have all been used in absorption spectrophotometry. Any reasonably stable light source with the necessary intensity of light at the needed wavelengths has the potential of being an adequate light source for absorption spectroscopy. Monochromatic light for the spectrophotometric assays can be obtained from the light source itself (e.g., lasers and light emitting diodes or from dispersed polychromatic light by the use of prisms and gratings and selected from polychromatic light by the use of interference and/or interference filters.[13] Most sample containers for absorption spectroscopy have a fixed path length [14] (typically 1 cm), and the volume of the sample will not affect the path length [15] as long as the level of the solution is higher than the light path. However, as was mentioned with the increasingly popular microtiter plates (or microplates), the sample volume will affect the path length. No matter which type of sample cell is used, the analyst must be certain that the material of the sample cell will be essentially transparent to the wavelength of the light being used in the assay. If the assay requires the use of wavelengths from 200 to about 350 nm, then the expensive quartz sample cells are needed. If the assay requires wavelengths from about 350 to 750 nm, less expensive glass or plastic cells can be used, although the expensive quartz sample cells can still be used. Both glass and plastic sample cells have some absorption from 350 to 400 nm, and their use for assays in this range of wavelengths is limited.

In modern spectrophotometric instrumentation, the transmitted light is detected with a detector, where it is converted to an electronic signal, which is subsequently amplified and converted into an output signal. All electronic spectrophotometers have some electronic signal even when no light reaches the detector (sometimes called a dark current). Thus each instrument must be calibrated to determine what electronic signal is due to the light and what

---

[13] Technically speaking, if filters or lasers or LEDs are used, then the process would be called absorption photometry, since the term *spectrophotometry* is generally considered to be limited to those systems with multiple wavelength capabilities. However, for ease of discussion we will use the term spectrophotometry to cover both photometry and spectrophotometry.

[14] Standard cuvettes vary from 0.1 to 10 cm path length.

[15] It is assumed in this discussion that the sample volume is sufficient to fill the sample cell so that all the light passes through the sample solution and that none passes over it.

electronic signal is due to the dark current.[16] This calibration is done by blocking the light path and measuring the electronic signal condition at zero light (0% T — infinite absorbance) and then by placing a sample cell with the blank solution in the sample compartment and measuring the electronic signal at 100% light (100% T — zero absorbance):

In single-beam spectrophotometry the sample cell and the reference cell are moved in and out of the light path. The operation of the single-beam spectrophotometer consists of first determining the dark current and then placing the reference cell filled with the appropriate blank (see below) in the light path and adjusting the light intensity such that an appropriate intensity is measured by the detector. This operation defines zero absorbance (i.e., the 100% T) reading on the spectrophotometer. Then the sample cell is moved into the light path, the light intensity is measured, and this light intensity is compared with that of the blank, and the ratio of these intensities is converted into an absorbance (see Eqs. 3.3 and 3.4).

In double-beam spectrophotometry both the sample cell and the reference cell are in the light path and the 100% light condition is determined continuously.[17] This is accomplished either by splitting the light beam or by chopping it. The intensity of the light passing through the sample cell is automatically compared with the intensity of the light passing through the reference cell. The use of a double-beam spectrophotometer significantly eases the compensation for changes in the original light intensity of the light source with time and/or wavelength. Double-beam spectrophotometers are the traditional instruments of choice[18] for measuring absorbance spectra; however, they are more expensive than single-beam instruments. Optical blanking (see below) may be used with both the traditional single-beam and double-beam spectrophotometers, but it is not normally used with microtiter-plate systems.

SPECTROPHOTOMETRIC ASSAYS

When doing spectrophotometric measurements, the analyst should first determine the linear range for the specific spectrophotometric analysis being planned. The process of determining the linear range for an assay is described in Appendix 3. Beer's law is only valid within the linear range of the assay. Once the linear range of the assay is determined, the concentrations of the analyte in solutions of unknown content are determined by measuring their absorbencies and

---

[16] Note that this calibration of the spectrophotometer is crucial to good spectrophotometric measurements. As was noted in Chapter 1, such calibration procedures are frequently crucial steps in the measurement process.

[17] The dark current must still be determined by some method.

[18] The diode array spectrophotometer will probably be the instrument of choice in the future.

mathematically computing the concentration (see Appendix 3). Since graphical computations of concentrations by extrapolation above the linear range are rather imprecise, the best strategy is to dilute the sample so that the concentration of the analyte is within the linear range, and then reanalyze the diluted sample.

ABSORPTIVITIES

If the absorbance measurements were made with a 1 cm path length and the concentration units were molar, then the slope of the linear portion of the standard curve is molar absorptivity (ε), which has the units of M^{-1} cm^{-1}. Molar absorptivity is usually given for those wavelengths at which there is an absorbance peak. For instance, a molar absorptivity for unprotonated bromcresol green would normally be reported for 616 nm (see Fig. 3.3), since that is the wavelength of its maximum absorbance. However, a molar absorptivity may be determined at any wavelength at which a compound absorbs light. However, the reader will remember that most compounds absorb different amounts of light at different wavelengths. Thus, when reporting a molar absorptivity, it is important to specify the wavelength.

Standard curves with different figures of merit[19] are obtained with absorbance measurements at different wavelengths for the same analyte. For example, plotting the absorbance at 616, 640, and 550 nm of a bromcresol green solutions (pH 10) versus the concentration of bromcresol green (Fig. 3.8) results in three straight lines over the concentration range used. The slopes of these lines and thus the molar absorptivity vary as a function of the wavelength. The largest molar absorptivity is always obtained at the absorbance wavelength maximum (λ_{max}). This wavelength (λ_{max}) is commonly used for estimation of the concentration of compound because spectrophotometric assays are most sensitive at this wavelength. On the other hand, if the analyte concentrations are high enough that the absorbencies at the λ_{max} are above the linear range, development of standard curves at wavelengths with lower molar absorbitivities may well result in a linear standard curve for these higher concentrations.[20]

ADDITIVITY

Since quantum levels of individual molecules are the determining factors in the absorption of light, the absorbance of light by a solution with multiple

---

[19] See Appendix 3.

[20] Assuming that the nonlinearity is due to stray light. See discussion later in this chapter on chemical reasons for nonlinearity.

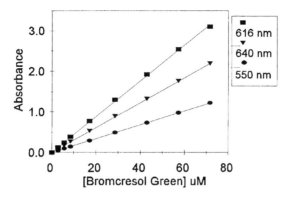

Fig. 3.8. Bromcresol Green (unprotonated form) standard curves at different wavelengths.

compounds each with their own absorbance of light is the sum of the individual absorbencies.

$$A_{(\text{mixture})} = A_{(\text{compound A})} + A_{(\text{compound B})} \cdots A_{(\text{compound N})} \qquad (3.8a)$$

This is true as long as there is no chemical interaction between any of the compounds. For biological analysis, this additivity is a crucial feature of absorption spectroscopy. Because if the absorbencies are additive, then the absorbance of an individual chemical species can be determined by subtracting the absorbance of all other chemical species from the total absorbance of the mixture.

$$A_{(\text{compound A})} = A_{(\text{mixture})} - \sum (A_{(\text{compound B})} \cdots A_{(\text{compound N})}) \qquad (3.8b)$$

Thus, although the total composition of a biological mixture may not known, the absorbance of a particular species can be determined by subtraction of the absorbencies of all the other species from the total absorbance. For example, if an analyte of a given concentration was added to the mixture shown in Eq. 3.8a, then the absorbance would be the sum of all the absorbencies. However, as shown further in Eq. 3.8b, subtracting the absorbance of the mixture without the analyte would yield the absorbance of only the analyte. This feature of absorption spectroscopy makes it a powerful tool in the chemical analyses of biological systems since there are very few other detection techniques that will permit such ready subtraction of background signals. The subtraction of background signals will be discussed in more detail later in this chapter under the discussion of blanks.

DEVIATIONS FROM BEER'S LAW

There are several key assumptions inherent in the use of Beer's law for quantitative measurements. They are that:

1. All absorbance measurements are made on clear[21] solutions (no turbidity).
2. All detected light has passed through the sample solution (no stray light).
3. Each absorbing species has an independent absorption of light, and there is no chemical interaction between those components in solution (no chemical interaction).
4. The output of the light source and the response of the detector are stable over the period of the absorbance measurements (no noise).
5. There is no loss of energy from the electronic excited state through the emission of the light (no fluorescence).

TURBID SOLUTIONS

Solutions are turbid because they contain finely dispersed multiple phases with different refractive indices. Such finely dispersed systems scatter light. Multiple-phase systems may be gas in liquid systems, immiscible liquids in liquid systems, and solids in liquid systems or some combination of these situations. All multiple-phase systems will introduce some turbidity and thus scatter the transmitted light. Since the scattered light does not reach the spectrophotometer detector, turbidity leads to an erroneous increase in apparent absorbance. The amount of light scattering depends upon the number and size of the particles and upon the wavelength of the light being used. Shorter wavelengths (higher frequency) of light are scattered more than longer wavelengths. There are specialized analytical applications of the light scattering by particles, and they will be discussed later in this chapter under the heading of nephelometry and turbidity.

STRAY LIGHT

The spectrophotometer detector responds to light whatever its source. However, if the light reaching the detector has not passed through the sample solution, the observed absorbance will be erroneously low. The magnitude of the error is:

$$A = -\log\left(\frac{(I_{fnl} + I_{stray})}{(I_0 + I_{stray})}\right) \qquad (3.9)$$

---

[21] Clear does not mean colorless; it denotes a solution that is not turbid (cloudy).

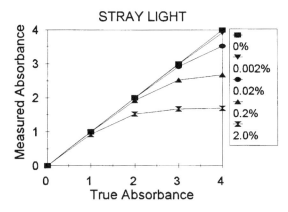

Fig. 3.9. Effect of stray light on linearity.

where: I_{fnl} is the final intensity of light, I_0 the initial intensity of light, and I_{stray} the intensity of detected stray light.

The impact of stray light is greater, the higher the absorbance of the solution. The net effect of stray light is to introduce nonlinearity at the higher concentrations; as shown in Figure 3.9. Increasing the amount of stray light in a spectroscopic measurement systems decreases the linear range of the response curve.

There are two primary causes of stray light: operator error and spectrophotometer design. Common operator errors that lead to stray light include failure to close the cuvette compartment of the spectrophotometer and failure to fill the spectrophotometer cuvette so that the entire light beam passes through the sample. If less than this volume is used, part of the light beam will pass above the sample, but still fall on the detector, that is, stray light. Poor spectrophotometer design can lead to stray light by failure to eliminate the detection of light that originates outside the instrument and/or light being reflected[22] around the sample cell and/or by failure to use monochromatic light in the spectrophotometer. Stray light originating from outside the spectrophotometer and/or by reflection within the instrument is a characteristic of the instrument and is normally provided in the instrument technical specifications. This value is presented in terms of percent transmission (%T) and is usually associated with a particular wavelength. A lower stray light value is indicative of a higher-quality instrument and usually translates to a higher price. Many inexpensive instruments have stray components on the order of 0.2%T, and very few instruments have less than 0.002%T stray light. Stray light resulting from imperfect selection of monochromatic light can occur if the harmonics[23] of the selected wavelength have significant intensity and the

---

[22] The reason that the optical compartments of spectrophotometers are painted flat black is to reduce the amount of stray light.

[23] This phenomenon is similar to sound from a stringed instrument that contains the primary note and harmonics.

absorbing components have different molar absorptivities at the harmonic wavelengths. Some spectrophotometers use optical filters to eliminate these harmonics.

It is difficult and expensive to produce a truly monochromatic light for a spectrophotometer, and thus almost all spectrophotometers utilize a polychromatic light source in which the intensity of the light is symmetrically distributed around the selected wavelength. In a number of relatively rare cases, if the bandwidth of the polychromatic light spans two or more absorption maxima of the analyte, the resulting nonlinearity will be similar to that observed with stray light. Experimentally such situations are rare and can be avoided if the bandwidth is narrowed or if the chemical matrix is changed so that the distinct absorption maxima are eliminated.

CHEMICAL INTERACTIONS

Many compounds undergo concentration-dependent association/disassociation (monomer–nmer) reactions in which the absorbance spectrum of the dissociated species is different from that of the aggregated species. In such cases a plot of absorbance versus concentration will become nonlinear at higher concentrations. Furthermore, most compounds show electrostatic interactions at concentrations of about $0.01 \, M$[24] which also lead to nonlinearity of the absorbance versus concentration. Other types of interactions that can cause problems include acid–base reactions between components of the solution, changes in local solvent conditions due to hydrophobic/hydrophilic reactions, ionic strength changes, and changes in the refractive index of the solutions. For example, chromate (CrO_4^{2-}) reacts with water to form $HCrO_4^-$. The ratio of these species in aqueous solution varies with dilution and their absorbance spectra differ. Thus there is not a linear relationship between the absorbance and the total concentration of chromate ($CrO_4^{2-} + HCrO_4^-$) in aqueous solutions. However, if the these solutions are prepared in $50 \, mM$ KOH to eliminate the formation of $HCrO_4^-$, chromate (or dichromate, $K_2Cr_2O_7$) solutions can be used as primary spectroscopic standards. Another example of a dissociation/association reaction involves $CuCl_2$, which is green in concentrated solutions. When $CuCl_2$ is diluted, a larger fraction of this salt dissociates to produce Cu^{2+} and a resulting blue solution. The relationship between absorbance and concentration for $CuCl_2$ would not be linear whether absorbance was monitored at the absorbance maximum of $CuCl_2$ or of Cu^{2+}.

It is not infrequent that the various components of the matrix undergo reactions with each other, the products of such reactions can also create chemical interferences. For example, several high molecular weight compounds alter their

---

[24] Some interactions occur at much lower concentrations; e.g., methylene blue is reported to undergo electrostatic interactions, which affect its spectrum at concentrations around $10^{-5} \, M$.

light scattering characteristics after interaction with heavy metals, detergents, heat, etc.

CHEMICAL VERSUS INSTRUMENT NONLINEARITY

It is possible to differentiate between chemical and instrumental causes of a nonlinear relationship between absorbance and concentration. If the behavior is the result of a chemical effect, the nonlinear relationship between absorbance and concentration should be observed at any wavelength suitable for measuring the concentration of the compound of interest. However, if the nonlinearity is an instrumental effect, it will only be observed when the absorbance is above a value characteristic of the instrument; that is, the nonlinear behavior will be a function of the wavelength used to make the observations. For example, consider the relationship between absorbance and the concentration of unprotonated bromcresol green. The absorbance maximum (λ_{max}) of unprotonated bromcresol is 616 nm (Fig. 3.3, pH 10). Absorbance is a linear function of concentration up to at least 75 µM bromcresol green (Fig. 3.10) regardless of wavelength. However, when the concentration is above 100 µM, the absorbance at 616 nm does not increase as a function of concentration (Fig. 3.10). If this phenomenon was the result of an association/dissociation reaction or aggregation, it should be observed at other wavelengths absorbed by bromcresol green. However, absorbance at 550 nm is a linear function of concentration up to at least 150 µM bromcresol green, and at 640 nm this relationship becomes nonlinear at about 120 µM bromcresol green (Fig. 3.10). The nonlinearity is observed when the absorbance of the sample exceeds ~3.0, a value reached at lower concentration when 616 nm (λ_{max}) is used to make the observation than when

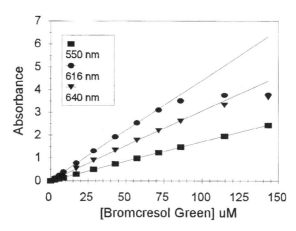

Fig. 3.10. Response curves for bromcresol green (unprotonated form) at different wavelengths.

640 nm is used. At 550 nm, this value is not reached and the absorbance/ concentration relationship is linear. However, if higher concentrations of bromcresol green were used, one should expect that this relationship would become nonlinear as the absorbance value exceeded ~3.0. The stray light specification of the instrument used to collect these data was 0.02%, a value consistent with the observation that nonlinearity was observed at absorbance values above 3.0.

NOISE AND OTHER SOURCES OF ERROR

All instruments have some noise. The output of the light source for a spectrophotometer will have a component of noise, and the response of the spectrophotometric detector will have one noise component at high light intensities and another at low light intensities. As shown in Table 3.2, the level of the light and detector noise will affect the level of the absorbance noise. Thus at lower concentrations of analyte, there will be some concentration where the absorbance of the analyte will be indistinguishable from the absorbance due to the noise of the light source and detector. Different spectrophotometers will have different amounts of noise due to the light source and the detector, and these noise levels usually change with the age and use of the instrument.

With those spectrophotometric instruments that have variable wavelength selection, there can be errors in the exact selection of the measurement wavelength. If the measurement wavelength is near a flat portion of the absorption spectrum (normally at the λ_{max}), then the resulting absorbance error should be relatively small. However, if the measurement wavelength is in a region where the absorption spectrum shows significant change with wavelength, then the absorbance error that will result can be rather large. The use of filter spectrophotometers will essentially eliminate this source of error. Given the

TABLE 3.2. The effect of Light and Detector Noise on the Absorbance Noise

Light and Detector Noise		Absorbance Noise
1 part in	10^6	0.000000434
1 part in	10^5	0.00000434
1 part in	10^4	0.0000434
1 part in	10^3	0.000434
1 part in	10^2	0.00436
2 parts in	10^2	0.00877
3 parts in	10^2	0.0132
4 parts in	10^2	0.0177
5 parts in	10^2	0.0223

multiple causes of nonlinearity of absorbance measurements, it is the better part of wisdom for an analyst to determine the linear range of each spectrophotometric assay. The process for determining the linear range is given in Appendix 3. In most cases the linear range for spectrophotometry is limited to a range of 0.01–1.5 AU; however, the linear ranges of individual spectrophotometers and assay systems can vary.

DILUTIONS

As noted, if the absorbance of the sample of unknown concentration exceeds the linear range, the sample should be diluted such that its absorbance is within the linear range. This absorbance now represents the concentration of the compound in the diluted solution, and the concentration of the compound in the original solution is the concentration in the diluted solution times the dilution factor as shown in Eq. 3.10.

$$[\text{concentration}]_{\text{original}} = [\text{concentration}]_{\text{diluted}} \times (\text{dilution factor}) \qquad (3.10)$$

For example, if the net absorbance was 3.660 for the analysis of a sample of unknown concentration with an analyte of molar absorptivity 18.3 mM^{-1} $cm^{-1} \times$ 1 cm, it would *not* be correct to calculate the concentration as 0.2 mM [$3.660/(18.3$ mM^{-1} $cm^{-1} \times 1$ cm)]. Most spectrophotometric assays are linear only to 2.500, and it is extremely unlikely that there was a linear response to 3.660 absorbance. In order to accurately determine the concentration of the compound, the solution should be diluted. For instance, assume that a tenfold dilution [25] of the original solution has an absorbance of 0.915. The concentration of this diluted solution is 0.05 mM [$0.915/(18.3$ mM^{-1} $cm^{-1} \times 1$ cm)], and the concentration of the original undiluted solution is 0.5 mM (0.05 mM × 10). Alternatively one could shorten the light path, and if the range for this assay system was still linear to 2.5 absorbance, then if the light path length was shortened to 0.2 cm, the absorbance of the solution would be 1.830 and again the concentration of this solution would be 0.5 mM [$1.830/(18.3$ mM^{-1} $cm^{-1} \times 0.2$ cm)]. Another alternative would be to use another wavelength at which the compound absorbs. A new molar absorptivity and linear range would have to be determined for this new wavelength, but otherwise the process is the same. Each of these three approaches (dilution, shortened light path length, and use of a wavelength other than λ_{max}) is designed to ensure that the absorbance value used

---

[25] A dilution factor of 10. One part concentrated solution and nine parts diluent. The concentration of the diluted sample is one-tenth that of the original sample.

to calculate the analyte concentration is within the linear range of the assay system.[26]

MIXTURES AND BLANKS

As was mentioned earlier, the absorbance of a solution with multiple compounds, each with its own absorbance of light, is the sum of the individual absorbencies. Furthermore, the absorbance of an individual chemical species can be determined by subtracting the absorbance of the other chemical species from the total absorbance of the mixture. These phenomena have led to the extensive use of blanks in the use of absorption spectrophotometry for the analysis of biological samples. Blanks are defined as solutions that contain everything present in an assay mixture except the analyte. The use of blanks can be readily shown by an example.

Example: Suppose you need to measure the spectrum of *p*-nitrophenolate (a common product of enzymatic assays) in the presence of beet juice in a buffer. You know neither the complete chemical composition of the beet juice nor the concentration of the components of the beet juice. Still determining the spectrum of *p*-nitrophenol in the beet juice is straightforward. The approach is outlined in the following. The following reagents are available to you:

> 50 mM glycine buffer, pH 10.0
>
> 0.01 g/L *p*-nitrophenol in water (pNP)
>
> beet juice (bj) diluted 1:25

and you prepare 4 mL of each of the solutions given in Table 3.3. You then measure the spectra of samples 2, 3, and 4 using sample 1 in the reference cell. (As you will remember, the reference cell is used to set the 0 absorbance reading on the spectrophotometer.) Your data are given in Table 3.4. You plot the absorbance values versus the wavelengths. The resulting plots are shown in Figure 3.11. The spectrum of sample 1 is zero at all wavelengths, since sample 1 was used as the blank. Now, if you repeat the determination of the spectra, but use sample 3 in the reference cell (i.e., sample 3 is the blank), then the spectrum of sample 3 will be zero at all wavelengths and the spectrum of sample 4 will be

---

[26] As an aside, it should be noted that absorbance is not a function of sample volume or analyte amounts but only of analyte concentration and length of light path. A 1 mM solution of a compound that has a molar absorptivity of 2 mM^{-1} cm^{-1} at a specific wavelength will have an absorbance of 2.000 at this wavelength (assuming a light path length of 1 cm) whether one has 1 mL or 1 L of the solution. Obviously, the amount of the compound is a function of the volume, but the concentrations and thus the absorbance values are identical.

TABLE 3.3. Solutions Used in Beet Juice Experiment

Sample No.	p-NP (mL)	Beet Juice (mL)	Glycine Buffer (mL)	Water (mL)	Total Volume (mL)
1			2	2	4
2	1		2	1	4
3		1	2	1	4
4	1	1	2		4

TABLE 3.4. Spectra of Solutions of Table 3.3 when Sample 1 Is Used as a Blank

	Sample No.			
Wavelength	1	2	3	4
330	0	0.07	0.191	0.269
350	0	0.151	0.186	0.344
370	0	0.287	0.202	0.497
390	0	0.404	0.188	0.601
410	0	0.39	0.149	0.541
430	0	0.223	0.109	0.332
450	0	0.078	0.092	0.173
470	0	0.015	0.09	0.108
490	0	0.002	0.109	0.113
510	0	0	0.13	0.134
530	0	0	0.145	0.15
550	0	0	0.137	0.142
570	0	0.002	0.113	0.117
590	0	0.001	0.084	0.087

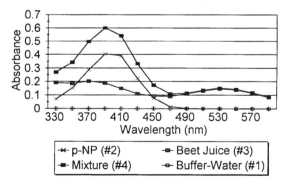

Additivity of Spectra
Water-Buffer Blank

Legend: p-NP (#2), Beet Juice (#3), Mixture (#4), Buffer-Water (#1)

Fig. 3.11. Experimental spectra: reference cell contains water/buffer.

identical to the spectrum of sample 2 when the water–buffer blank was used to measure its spectrum. The use of sample 3 as a blank is an example of optical subtraction.

Alternatively, you could have done a mathematical subtraction of the spectrum of the diluted beet juice on a wavelength-by-wavelength basis and you would have still noted that the spectrum of sample 3 will be zero at all wavelengths and the spectrum of sample 4 will be identical to the spectrum of sample 2 when the water–buffer blank was used to measure its spectrum. This is an example of mathematical subtraction. The optical subtraction method is usually preferred since it is easier to do and because the use of the blank to set the zero absorbance results in less interference from stray light. Remember that the subtraction of blanks can only be used when there is no interaction between the components of the mixture. If the spectrum of one of the compounds obtained by subtraction is different from that obtained in pure solution, that is good evidence that a chemical reaction of some type has occurred.

TURBIDIMETRY AND NEPHELOMETRY

When a sample has uniform-sized particles suspended in solution, then light scattering can be used to determine the concentration of these particles. Turbidimetry designates the use of light scattering when the observed light passes directly though the turbid solution; that is, the transmitted and forward scattered light is measured. Nephelometry designates the use of light scattering when the scattered light is measured at some angle to the irradiating light. Generally 90° angles are used, although angles varying from 75° to 135° have been used. When turbidimetry is used, the equation relating the scattered light to concentration is similar to Beer's law.

$$\tau = -\left(\frac{2.303}{b}\right)\log\left(\frac{I_{Fnl}}{I_0}\right) = a'bc \tag{3.11}$$

where I_{Fnl} and I_0 are the intensity of the transmitted light and the intensity of the original transmitted light, respectively; τ is the turbidity, a' is the turbidimetric absorptivity, b is the path length, and c is the particle concentration. Note that a' is a function of the particle size, the wavelength of the scattered light, and the instrumentation.

When nephelometry is used the intensity of the scattered light is directly proportional to the particle concentration:

$$I_{Fnl} = k'c \tag{3.12}$$

where I_{Fnl} is the intensity of the scattered light, c is the particle concentration, and k' is a constant that is a function of the particle size, the wavelength of the scattered light, and the instrumentation.

Note that a' is a function of the particle size, the wavelength of the scattered light, and the instrumentation. The amount of light scattering is proportional to the size and shape of the particles, as well as the wavelength of the incident light and the design of the instrumentation. Calibration of individual assays is necessary. It is crucial that the wavelength of the light used to make the measurement is not one that is absorbed by any of the components of the solution; 600 nm is frequently used. Light scattering is frequently used to estimate the concentration of bacteria in liquid cultures. Since different bacteria have different sizes and thus have different light scattering properties, one must have a secondary method (plate counts) to establish the relationship between light scattering and cell density in liquid culture for any particular bacterium.

ATOMIC ABSORPTION SPECTROPHOTOMETRY

Atomic absorption spectroscopy can be used for the assay of up to 70 elements. This form of absorption spectroscopy utilizes the formation of atomic gas followed by the measurement of the absorption of selected light by the ionized elemental atoms. Atomic absorption spectra are line spectra (see Fig. 3.5), and multielement samples can have very complex, overlapping spectra. This analytical technique requires the digestion of the biological sample to destroy the various matrix compounds; the digested samples are then atomized by very high temperatures (1700–10,000°C), utilizing flames, furnaces, or plasmas. The light sources are hollow-cathode lamps or electrodeless discharge lamps. High-resolution spectrophotometers are utilized to resolve the complex spectra generated by multielement samples. Atomic absorption spectroscopy is routinely used for the analysis of the trace metal content of biological samples because it has the necessary sensitivity and selectivity for these assays. Its primary disadvantages are the high cost of the instrumentation, the need for the extensive and careful sample digestion, and the loss of information on the organic chemical structures that contained the elements in the biological samples. Interferences that lead to errors are common in mineral and trace element analyses, and good analysts routinely include standard reference materials with known mineral and trace element compositions in their analysis sets to validate their analytical work.

FLUORESCENCE SPECTROSCOPY

As was mentioned previously, once an electron is excited to a higher-energy state, then that electron can lose energy through emission of a photon. If the electron was excited to the higher-energy state by a chemical reaction, then the emission process is called *luminescence.*[27] If the electron was excited to the higher-energy state by the absorption of light, then the emission process is called *fluorescence.* A fluorescent molecule has both an absorption spectrum and an emission spectrum. The emission spectrum always has longer wavelengths (lower energies) than the absorption spectrum (see Fig. 3.12). Fluorescence spectrophotometers have the general configuration shown in Figure 3.13. Unfortunately, many compounds that have absorption spectra do not fluoresce, but lose excitation energy through molecular vibrations, motions, and collisions with other molecules. However, if a compound does fluoresce, its fluorescence intensity can be significant.

Fluoresence assays are often several orders of magnitude more sensitive than absorption spectroscopy assays of the same compound. This increase in sensitivity can be of significant use in the analyses of the levels of analytes with concentrations in the nanomolar to picomolar concentrations. The amount of light emitted by a light-absorbing molecule is directly proportional to its concentration:

$$F = Kc \tag{3.13}$$

where F is the fluorescence intensity, K is a constant for the instrument and excitation energy, and c is the concentration of the fluorescent species.

Unfortunately, there is a significant variation in the fluorescent intensity with different light sources, different instrument geometries, and different detector responses, and it is necessary to calibrate each spectrofluorometer prior to each

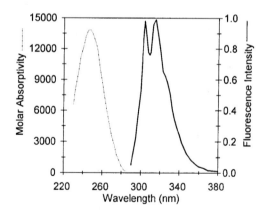

Fig. 3.12.   The absorption and emission spectra of bi-phenyl.

---

[27] The use of luminescence in chemical measurements in biological systems is discussed in Chapter 6.

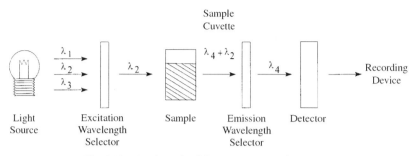

Fig. 3.13. A schematic of fluorescent spectrochotometer.

set of analytical measurements. Spectrofluorometers are usually calibrated with compounds such as quinine or fluorescein. While fluorescence standard curves are linear at low concentrations (absorbencies less than 0.04), they frequently become nonlinear at higher concentrations Interferences in biological samples due to the fluorescence of other compounds are not uncommon, and the analyst needs to run the appropriate controls and blanks to ensure that the measured fluorescence is due to the analyte under study.

CHEMILUMINESCENCE

There are a number of reactions that emit photons of light as one of the products of the reaction. Such reactions can yield very sensitive assays and will be discussed in Chapter 6.

4. Detection Reactions (Colorimetric Reactions)

INTRODUCTION

The principles of quantitative spectrophotometric analysis developed in the previous chapter involved direct measurement of light-absorbing compounds. However, many of the analytes of interest to the life sciences analysts do *not* have chromophores that absorb light in a useful part of the ultraviolet or visible range. Thus spectrophotometry cannot be directly used to determine the concentration of these compounds. However, there are a number of specific chemical reactions that transform these analytes into colored products that do have adequate absorbance.[1] These detection reactions are frequently called derivatization or colorimetric reactions. Typical spectrophotometric changes that occur in colorimetric reactions are shown in Figure 4.1. In the ideal colorimetric assay the absorbance of the product is proportional to the concentration of the original colorless analyte. In reality the absorbance of the product is proportional to the concentration of the original analyte only under certain defined reaction conditions, within restricted analyte and reagent concentration ranges, and within given time spans.

---

[1] While most this chapter makes specific reference to absorption spectroscopy, the principles are generally applicable to the colorimetric reactions which produce fluorescent or chemi-luminescent products.

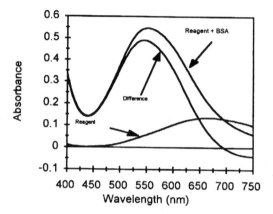

Fig. 4.1. Spectra of Biuret assays. The two spectra labeled reagent and reagent + BSA used water as a blank, the spectrum labeled difference used a reagent blank. (See below for details).

REACTION TIME COURSE

A typical time course for the development of chromophores is shown in Figure 4.2. In the time course of a reaction the concentrations of the analyte and of the product are functions of time in the kinetic region and are independent of time in the equilibrium region. Sometimes the detectable products are themselves susceptible to degradation processes. Generally, the effect of the degradation

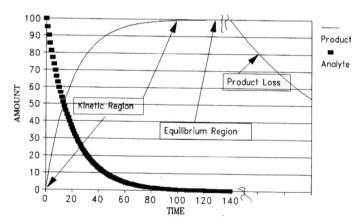

Fig. 4.2. Time course of a reaction. Pseudo-first order kinetics may be used for determination of the analyte concentration in that portion of the graph where the rate of production of the product is not a function of time. (*i.e.* at v_0)

processes can be minimized by measuring the product as soon as equilibrium is attained, thus minimizing the time for the degradation process to take place.

Occasionally, the existence of the reaction product is transitory, and the system never reaches equilibrium. In such cases the measurement needs to be done at that time at which the product concentration is maximum. In some cases the detection process itself is the cause of the loss of the product. For example, the measured products in several of the fluorescence assays for amino acids are susceptible to photodecomposition.

There are two types of colorimetric assays: equilibrium and kinetic. In equilibrium assays the concentrations do not change during the time needed for the measurement process; in kinetic assays the concentrations may change during the measurement process. Equilibrium colorimetric assays are the most common, and they will be discussed first. The use of kinetic assays to determine reactant concentration will be discussed in the latter part of this chapter. In biological systems, the use of kinetic assays to measure catalyst concentrations is usually done for the measurement of enzyme activity, and that topic will be discussed in Chapter 5. The use of enzymes in colorimetric reactions will be discussed in Chapter 6.

EQUILIBRIUM ASSAYS

In equilibrium colorimetric assays the absorbance of the product is the measured component. Since time does not influence the magnitude of the signal, the product signal may be acquired at any time during the equilibrium portion of the reaction. When the product formed has an absorbance that is a linear function of the analyte concentration (i.e., Beer's Law applies), a molar absorptivity could be calculated and used for determining the concentrations of the unknowns. However, molar absorptivities are normally *not* used in colorimetric assays due to frequent interferences from sample matrices and fluctuations caused by changes in the reaction conditions. In almost all situations where colorimetric reactions are used, the determination of the concentration of the analyte in the unknown is achieved by the use of a standard curve.

STANDARD CURVES

When doing equilibrium colorimetric analyses, the standard curve for the analyte is prepared at the same time as the analysis of unknown samples. Use of a standard curve not only controls the inherent paranoia of all analysts but also eliminates errors arising from deterioration of specific reagents during prolonged storage and usually compensates for errors from inadequate minor timing and/or temperature variations (see the following). A standard curve is prepared using a "no-analyte blank," which contains the reagents and other components of the mixture to be analyzed (e.g., the buffer) but has "zero" analyte concentration in

TABLE 4.1. Determination of a Standard Curve[a]

Sample No.	Protein Conc. (mg/mL)	Sample Vol. (mL)	Biuret Vol. (mL)	ABS	Blank with No.
1	0	1	4	0.000	1
2	2	1	4	0.143	1
3	4	1	4	0.261	1
4	6	1	4	0.379	1
5	Unk 1	1	4	0.168	1

[a]The regression analysis yields: intercept $(b) = 0.0075$ AU; $R^2 = 0.9999$; slope $(m) = 0.0628$ AU/mg/mL. Unknown ABS $= 0.146$. Using the equation for a straight line, $y = mx + b$, then x the concentration of the analyte in the unknown $= (y - b)/m$, and therefore the concentration of the unknown $= 2.56$ mg/mL.

the sample volume, and a series of samples containing all the reagents and increasing concentrations of the analyte being analyzed. A typical reaction set for the determination of a standard curve for a colorimetric reaction is shown in Table 4.1. The samples are then reacted with the reagent and their absorbance determined. These data are then analyzed using regression analysis (see Appendix 3) and the figures of merit determined. Samples with unknown analyte concentrations are reacted and analyzed in the exact same fashion as the standards. In fact, best practices suggest that the unknowns should be analyzed at the same time and with the same reagents as the standards. The final analyte concentrations in these samples are determined using the equation for a straight line.[2] These analyses of standard samples of known concentrations are a type of *positive control*. Positive controls are run to assure the analyst that the assay is working as it should. The basic assumption in the use of standard curves is that the response factor of the analyte in the unknown sample is the same as the response factor of the analyte in standard samples. This assumption is not always warranted, and the analyst needs to utilize the appropriate assay quality-control procedures to validate individual colorimetric analyses.

CRITICAL COMPONENTS OF COLORIMETRIC ASSAYS

The critical components of equilibrium colorimetric reactions are: that the amount of the product formed is directly proportional to the amount of analyte in the sample; that the product has a high molar absorptivity; that the detection reaction is selective for the analyte; that the colorimetric reaction is quantitative; and that the reaction reaches equilibrium within a reasonable amount of time.

---

[2] There are a number of colorimetric assays that give nonlinear responses. The analyst can either find a sufficiently narrow concentration range of the analyte where the response curve is close to being linear, or use nonlinear regression analytic techniques. The use of nonlinear regression techniques will not be covered in this book.

The critical components of a colorimetric assays are:

1. Product concentration $\propto$ analyte concentration.
2. Product has a high molar absorptivity.
3. Detection reaction is selective for the analyte.
4. Reaction is 99%+ complete.
5. Reaction time is reasonable.

PROPORTIONALITY

The requirement that the amount of a product formed is directly proportional to the amount of analyte is fundamental to colorimetric assays. This requirement implies that none of the analyte is consumed by side reactions and that the reaction product is stable over the measurement period. Side reactions that consume the analyte and/or side reactions that destroy the colored product will usually result in a product concentration that is not proportional to the original concentration of the analyte. Many biological samples contain enzymes that will consume an analyte in a short period of time. However, detecting analyte loss can be difficult. Sometimes spiking (see Chapter 5) combined with time studies will permit detection of the problem, but even then it is rare that proper correction factors can be established. Usually the best that can be done is for the analyst to know the general chemical composition of the sample matrix and to use those assay reagents and protocols that limit the destruction of the analyte.

MOLAR ABSORPTIVITIES

Usually an analyst will want the colorimetric product to have a large molar absorptivity. Most products of colorimetric reactions have molar absorptivities in the range of $10^3 - 10^5$ AU/mole/liter. Theoretical computations suggest that the maximum possible for molar absorptivity is about 10^5 AU/mole/liter. Colorimetric products with molar absorptivities less than 10^3 are generally not acceptable, since such low molar absorptivities usually result in assays that lack adequate sensitivity and limits of detection.

REACTION SELECTIVITY

Understanding the selectivity of the detection reaction for the analyte is an important part of the selection of an assay method. Some colorimetric derivatizations are based upon chemical coupling with a specific functional group of the analytes, and some are based upon a specific reaction. Those chemical reactions based upon specific functional groups will normally react with *any* compound that has the same functional group. If the samples contain other components in their matrix that have some functional groups in common with the analyte, then the signal from the colorimetric reaction will be that due to *both* the analyte and the interfering compounds, and the value calculated for the analyte in the sample will be high and inaccurate. An example of such interferences would be the use of ninhydrin (which reacts with free α-amino groups) to determine total protein. A ninhydrin assay would measure all the proteins *and* all the other free amino groups that were present in the free amino acids and in the peptides in the sample solution (see Fig. 4.3).

Other assays are based upon changes to the reagent caused by reaction with the analyte (e.g., oxidation or reduction of the reagent). In such cases all compounds that can transform the reagent will give a change in total product signal. An example of such interferences would be the use of bicinchoninic acid (BCA) reagent (in which the Cu^{+2} is reduced to Cu^+, and the Cu^+ is then detected upon complexation with the BCA) to determine total protein. A BCA assay would measure all the proteins *and* all the other compounds that would reduce the cupric ion (e.g., reducing sugars; see Fig. 4.4).

Still, other assays are based upon the change of the solvent environment of the reagent chromophore. For example, some protein binding assays utilize the shift of the chromophore absorbance maximum wavelength and the change in absorbance that occurs when the chromophore moves from an aqueous to a nonaqueous environment. Nonanalyte matrix components that inadvertently

Assay Reaction
      Protein-NH$_2$ + Ninhydrin -----> Purple color

Interfering Reaction
      Amino Acid + Ninhydrin -----> Purple color

Fig. 4.3. The ninhydrin assay of proteins.

Reaction
      Protein + CuII ---> CuI
      CuI + BCA ---> Blue Color

Interfering Reaction
      Reducing Agent + CuII ---> CuI

      CuI + BCA ---> Blue Color

Fig. 4.4. The chemistry of the BCA assay.

modify the equilibrium of the hydrophobic nature of the reaction mixture in either direction may well give erroneous results.

QUANTITATIVE PRODUCT PRODUCTION

It is always desirable that the colorimetric reaction be quantitative (i.e., greater than 99% of the analyte has been converted into product). Lower than 99% conversion yields a less sensitive assay and may result in imprecise results. When the reaction is less than 99% complete, say, 85% complete, then minor fluctuations in reaction conditions (temperature, reaction time, reagent concentrations, etc.) can lead to 85% reaction in one case, 82% in another, and 87% in another. Incomplete reactions are usually caused by selection of inadequate time–temperature combinations and/or inadequate reagent:analyte ratios.

TEMPERATURE

The reaction temperature affects the time it will take a reaction to reach equilibrium (see Fig. 4.5) but usually has only a small effect on the product concentration at equilibrium in aqueous solutions. The rates of most chemical reactions are dependent upon the reaction temperature, and the rate of a reaction usually doubles for every 10 °C increase in the reaction temperature. Thus the analyst frequently has a tradeoff; higher than room temperature reactions mean shorter times to equilibrium but usually require the use of heated reaction baths; room temperature reaction temperatures mean longer times to equilibrium but do not require heating baths. The temperature dependence has two important ramifications for analysts doing equilibrium colorimetric reactions. First, too low a temperature may lead to incomplete reactions, which almost always lead to imprecise results. Second, when the derivatization reactions take too long, increasing the temperature will decrease the time required to reach reaction equilibrium.

Reagent Concentrations

A common cause of nonlinear colorimetric assays is an inadequate ratio of reagent concentration to analyte concentration. Under normal conditions, colorimetric reactions are done with a large excess of reagent. This is done for the following reasons:

1. Under the law of mass action,[3] the percentage of analyte converted to product is a function of both the reagent:analyte ratio and the equilibrium constant for the colorimetric reaction. Thus the law of mass action predicts that

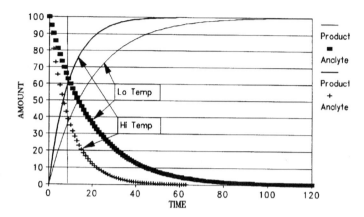

Fig. 4.5.   Effect of temperature on the reaction time course.

excess reagent will tend to ensure that more analyte will be converted to product. Usually a ratio of reagent to analyte concentration of 100 is adequate for quantitative (>99.9%) conversion of the analyte to product.

2. Fundamental kinetic theory states that the rates of all reactions are dependent upon the concentrations of all the components of the reaction. Thus higher reagent concentrations lead to faster reactions, which lead to quicker attainment of equilibrium.

3. When the zero analyte–reagent blank is used to zero a colorimeter, the assumption is that the reagent concentration does not significantly change during the colorimetric reaction. If the reagent concentration does significantly change during the reaction, then the basic assumption for blanking with the zero analyte–reagent blank is no longer valid (see below).

There are a number of cases where the reagent or some component of the reagent mixture is consumed by side reactions. Not infrequently these side reactions deplete the reagent concentration so that the assumption of excess reagent is not correct. For example, oxidation of oxygen-sensitive reagents is common. Excessively high analyte concentrations can deplete the reagent concentrations to less than the desired level.

---

[3] The law of mass action states that if you have a reaction where reactants A + B are in equilibrium with products C + D and the equilibrium constant is K_{eq}, then the concentration of [C][D] = [A][B]K_{eq} at chemical equilibrium.

Reaction Product Stability

While it seems intuitive that the reaction product needs to be stable over the measurement period, analysts sometimes overlook this important consideration. Knowledge of product stability is key in the selection of the time for the measurement of the amount of a product formed in a colorimetric reaction. If a product is stable for hours under the conditions of the analysis, then the absorbance measurements can be made almost at one's leisure; however, if the product has a very short lifetime (seconds or milliseconds), then the product concentrations must be measured rapidly to stay within the time window of product stability.

BLANKING AND SELECTION OF WAVELENGTH FOR DETECTION OF THE PRODUCT OF COLORIMETRIC REACTIONS

Generally an analyst is advised to select a wavelength for absorbance measurements where the product has a strong absorbance, where there is no overlap with the absorbance of other components in the mixture, and where small errors in wavelength settings do not lead to large errors in absorbencies. There are several likely possibilities for the overlap of the spectra of reagents, products, and matrices of samples. These possibilities are shown in Table 4.2.

No Overlap of Product Spectrum with Other Spectra

In case 1 of Table 4.2 (see Fig. 4.6) the measured absorbance at the product wavelength of maximum absorbance will be due only to the reaction product. Neither the sample matrix containing the analyte nor the reagent will have any absorbance at the detection wavelength. Thus there will be no interference by the reagent or matrix compounds. This is the best-case scenario for colorimetric reactions. Typically the analyst uses the wavelength where the product has a maximum absorbance.

TABLE 4.2. Does the absorption Spectrum Overlap the Absorption Spectrum of the Product?

Case Number	Reagent	Analyte	Product	Matrix of Sample	Example
1	No	No	Yes	No	Fig. 4.6
2	Yes	No	Yes	No	Fig. 4.7
3	Yes	No	Yes	Yes	Fig. 4.8
4	?	Yes	Yes	?	

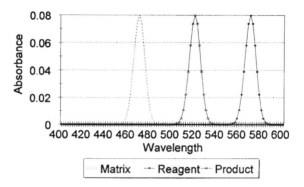

Fig. 4.6. Non-overlapping spectra.

THE PRODUCT AND THE REAGENT HAVE OVERLAPPING SPECTRA

In case 2 of Table 4.2 (see Fig. 4.7) the measured absorbance at the product wavelength of maximum absorbance will be due to the reaction product and to the reagent. The sample matrix does not contain any compound that has any absorbencies at those wavelengths where the product absorbs light. This is a common occurrence in colorimetric assays. Typically the analyst still uses the wavelength where the product has a maximum absorbance. However, when the product and reagent spectra overlap, then the absorbance due to the reagent must be subtracted from the total absorbance of the system so that the absorbance of the product can be determined. This absorbance subtraction is most easily done by using the reaction mixture that has a "0 analyte" blank as the reference solution for blanking the spectrophotometer. See Table 4.1 for an example of how this blanking is done. When the product and reagent spectra overlap, it is important that there is no substantial change in the concentration of the reagent during the reaction because the blank signal is subtracted from the final signal. If

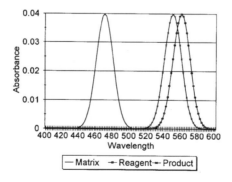

Fig. 4.7. Spectra overlap of reagent and product.

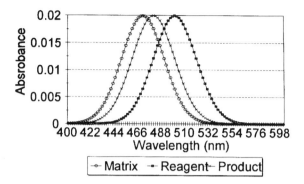

Fig. 4.8. Extensive spectra overlap.

the concentration of the reagent significantly decreases (>1% loss), then absorbance due to the reagent will change and the assumption underlying the use of blank subtraction may not be valid. In practice, maintaining the reagent concentration at 99% of its original concentration will usually be adequate. Such conditions are usually met by having the reagent concentration about 100-fold greater than the analyte concentration.

A more difficult case for blanking colorimetric analyses is that of case 3 in Table 4.2 (Fig. 4.8). The sample matrix, the product, and the reagent each have absorbencies at the detection wavelength for the product. In this case the total absorbance will be that of the sample matrix plus that of the solvent[4] plus that of the reagent plus that of the reaction product. Analysts must design their blanks so that the absorbencies due to the product can be determined and reported. The proper design of blanks in this case requires two sets of reaction mixtures, one that contains reagent and the other that contains only reagent solvent. The analyst then determines the total absorbance of the reaction system and then subtracts the absorbencies due to the reagent, the matrix, and the sample solvent. The following protocol will accomplish that goal.

Protocol

Step 1: Prepare a standard curve for the assay with known concentrations of analyte. Use the "0" analyte sample to zero the spectrophotometer. At the same time, the colorimetric analyses are done on each of the three samples containing the unknown levels of analyte. The total absorbance of the reacted samples containing the unknown levels of analyte is shown in the following:

$$ABS_{\text{total reaction mixture}} = ABS_{\text{reaction product}} + ABS_{\text{reagent}} + ABS_{\text{matrix}} \qquad (4.1)$$

---

[4] Given that the solvent for many colorimetric reactions is water, there is often no concern about solvent absorbance. The solvent absorbance is discussed here for completeness.

The optical blank that will be used to zero the spectrophotometer is the standard sample "0" added analyte, which contains a reagent concentration equal to that in the unknown samples, *but it will not have any analyte or sample matrix* present in the solution. Its total absorbance will be as shown:

$$ABS_{\text{total blank mixture}} = ABS_{\text{reagent}} \tag{4.2}$$

When the absorbance of the blank is optically subtracted from that of the total reaction mixture, the net absorbance for step 1 will be:

$$ABS_{\text{net step 1}} = ABS_{\text{product}} + ABS_{\text{matrix}} \tag{4.3}$$

Step 2: Determine the absorbencies due to the sample matrices in each of the three samples. Take the three separate samples with different levels of analyte and add an amount of reagent solvent[5] equal to the volume of reagent used in step 1 but *does not have any active reagent*. The absorbance of the matrix is then determined using an optical blank containing *only* sample solvent and reagent solvent. The net absorbance of these samples is as shown:

$$ABS_{\text{net step 2}} = ABS_{\text{matrix}} \tag{4.4}$$

Step 3: Subtract the absorbencies determined in step 2 from those obtained in step 1. The resulting absorbencies will be those due to the product of the detection reaction in the samples. The absorbance due only to the reaction product is as shown:

$$ABS_{\text{net step 1}} - ABS_{\text{net step 2}} = ABS_{\text{reaction product}} \tag{4.5}$$

Finally use the results from the regression analysis of the known standards to determine the concentration of the analyte in the samples with unknown analyte concentration in the standard manner.

Example: You are studying the fortification of cola drinks with protein. You use the Biuret assay to determine the protein in the cola drinks. The problem is that the cola drinks have caramelized sugar that has some absorbance at 550 nm, where the absorbance of the Biuret reaction product is measured. Question: How is the protein content of the cola drink measured in this situation?

---

[5] It is important to use the reagent and sample solvents in the measurements of the matrix absorbance so as to account for the effects these solvents may have on the absorbance of light by the chromophores in the matrix, such as effect of pH.

TABLE 4.3. Determination of a Standard Curve and Analysis of Unknowns[a]

Sample No.	Protein Conc. (mg/mL)	Sample Vol. (mL)	Biuret Vol. (mL)	ABS (Blanked with Sample A)
A	0	1	4	0.00
B	2	1	4	0.122
C	4	1	4	0.240
D	6	1	4	0.358
Unk 1	?	1	4	0.23
Unk 2	?	1	4	0.15
Unk 3	?	1	4	0.46

[a]The regression analysis yields: intercept (b) 0.0012 AU; $R^2 = 0.99993244$; slope (m) = 0.0596 AU/mg/mL.

The analyst needs to determine the total absorbance of the reaction system and then devise a protocol for subtracting the absorbencies due to the reagent, and the matrix. The following protocol will accomplish that goal.

Step 1: Prepare a standard curve for the Biuret assay with known concentrations of protein (bovine serum albumin). Use the "0" protein sample from Table 4.3 as a blank to zero the spectrophotometer.

Step 2: Analyze the cola samples for their protein concentrations using the Biuret assay. Use the "0" protein sample from step 1 as a blank to zero the spectrophotometer (see Table 4.3).

Step 3: Determine the absorbencies due to the caramelized sugar in the cola samples. Use the "0" cola sample as a blank to zero the spectrophotometer (see Table 4.4).

Step 4: Subtract the absorbencies determined in step 3 from those obtained in step 2. The resulting absorbencies will be those due to the product of the Biuret reagent with the protein in the cola drinks (See Table 4.5). Use the results from

TABLE 4.4. Determination of Unknown Matrix Absorbencies

Sample No.	Unk No.	Sample Vol (mL)	Biuret Vol. (mL)	Sample Solvent Vol. (mL)	Biuret Solvent Vol. (mL)	ABS	Blank with Sample No.
1	1	1	0	0	4	0.061	4
2	2	1	0	0	4	0.062	4
3	3	1	0	0	4	0.060	4
4	Blank	0	0	1	4	0	4

the regression analysis of step 1 to determine the concentration of the protein in the cola drinks:

$$\text{mg/mL protein} = (\text{net ABS}_{unk} - \text{intercept}_{regression})/\text{slope}_{regression} \qquad (4.6)$$

TABLE 4.5. Estimation of Protein Concentration

Unknown No.	ABS Step 2	ABS Step 3	Net ABS	Concentration Protein in Cola Drink (mg/mL)
1	0.23	0.061	0.169	2.82
2	0.15	0.062	0.088	1.46
3	0.36	0.062	0.298	5.0

THE PRODUCT AND THE ANALYTE HAVE OVERLAPPING SPECTRA

In case 4 of Table 4.2 the analyte and the product have overlapping spectra. This case is an extremely difficult one to unravel. While there are some approaches that can be taken in the development of the solutions for these problems, the process can be very difficult. While in the first approximation an analyst might be tempted to use blanking identical to the situation where reagent, matrix, and product have overlapping spectra, this is *not a good idea*, because reactions that produce colored product will probably alter the electronic structure of the analyte. Unfortunately, such reactions will also probably alter the electronic structure of the component that originally gave the analyte its overlapping structure. Thus the problem becomes what to use as a blank for the final color of the colored component of the analyte. This is not an easy problem to solve, and the analyst should consider finding alterative solutions. The most obvious solutions are: (1) find another assay wherein the spectra of the product and the analyte do not overlap; (2) find a way to remove the component part of the analyte that has overlapping spectra with the reaction product.

KINETIC COLORIMETRIC MEASUREMENTS

In the kinetic portion of the time course of an analysis the product concentration changes with time. Thus in kinetic assays the parameter measured is the change in product concentration with time (i.e., dP/dt). Classical chemistry kinetic theory states that the rate of formation of a product C by the reaction shown in Figure 4.9 is given by:

$$\frac{dC}{dt} = [A][B]k_1 \qquad (4.7)$$

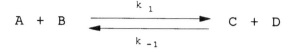

Fig. 4.9. Classical kinetic reactions.

where [A] and [B] are the concentrations of the reactants and k_1 is the rate constant for the reaction.

The rate of destruction of any of the product at any time ($-dC/dt$) is proportional to the concentration of the reactants (C, D) and the rate constant (k_{-1}):

$$-dC/dt = [C][D]k_{-1} \tag{4.8}$$

where [C] and [D] are the concentrations of the products and k_{-1} is the rate constant for the reaction.

Thus the *net rate* of formation of one compound at any time (dC/dt) is proportional to the rate of its formation minus the rate of its destruction:

$$\frac{dC}{dt} = [A][B]k_1 - [C][D]k_{-1} \tag{4.9}$$

If the reaction conditions are such that the reagent concentration is much, much greater than that of the analyte and the rate is measured at a time very early in the reaction time course where there is essentially no product to be destroyed, then we can rewrite Eq. 4.9:

$$\frac{dC}{dt} = [A]k_1' \tag{4.10}$$

where [A] is the concentration of the analyte and k_1' is equal to the concentration of the reagent [B] multiplied by the rate constant k_1.

This equation is a *pseudo-first-order* rate equation. With pseudo-first-order rate equations the rate of the reaction is proportional to the concentration of the analyte. Thus an analyst can use the rate of formation of a product and pseudo-first-order rate equations to determine the concentration of an analyte.[6] Kinetic assays using pseudo-first-order kinetics can be used with most colorimetric reactions if the measurements are made early enough in the time course of the reaction, that is, before there is any reverse reaction, and if the ratio of the reagent concentration is about 100-fold greater than the analyte concentration. The product signal is acquired either at a fixed time or as the slope of the signal as a function of time. If the measurements are made at fixed times, the response curve will be a plot of the absorbance at the fixed time versus the concentration of the

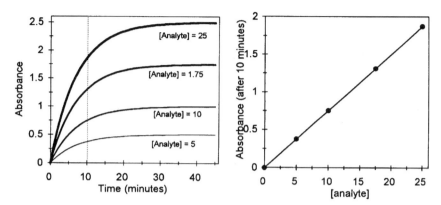

Fig. 4.10   Determination of reaction rates using fixed time data acquistion.

analyte (see Fig. 4.10). If the measurements are the slope of the concentration change, the response curve will be a plot of the slope versus analyte concentration (see Fig. 4.11). Those cases where the time course of the reaction has a lag phase (see Fig. 4.12), the rate measurements be made on the linear portion of the product concentration versus time plot. Fixed-time kinetic assays do have the advantages of ease of use and lower equipment cost since they do not require recorders or similar equipment. However, since the product signal is not recorded at multiple times, the analyst is not able directly to determine if the assumption of a fixed rate with time is correct. The slope method has the advantage that, with multiple signal/time data points, the analyst can check to see if the change of concentration with time is linear. Furthermore, the larger number of data points collected reduces the effect of noise on observations. The slope method has the disadvantages of more complex computations and higher instrumentation costs. Kinetic assays are more subject to interferences than equilibrium assays, and thus it is important that the analyst utilize the appropriate quality-control measures.

EFFECT OF TEMPERATURE ON KINETIC ASSAYS

Temperature can have a significant effect on kinetic measurements. There are direct effects on the reaction rates as defined by the Arrhenius equation, and there are indirect effects of temperature on reaction rates, such as the change in the physical structure of an enzyme due to its denaturation.

---

[6] This is one type of kinetic assay. The other type of kinetic analysis is used to determine the concentrations of catalysts. The most common assay for the determination of catalyst concentration in biological systems is the determination of enzyme activity, and this topic will be discussed in Chapter 5.

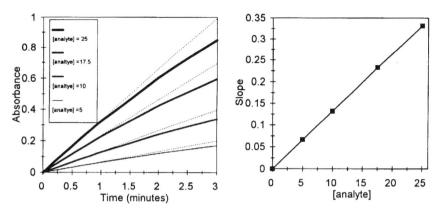

Fig. 4.11. Determination of reaction rates using slope measurements.

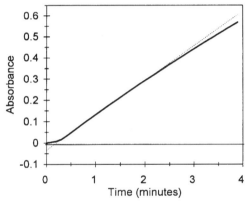

Fig. 4.12. Time course of a reaction with alag phase. Pseudo-first-order kinetics may be used for determination of the analyte concentration in that portion of the graph where the rate of production of the product is constant. (*i.e.*, after lag phase and before the plot of product versus time becomes non-linear)..

The Arrhenius equation states that if you have a reaction where A + B → C + D with a rate constant k, then the effect of temperature on the rate constant k is:

$$\log_{10} k = \frac{-E_a}{2.303RT} \qquad (4.11)$$

where E_a is the activation energy of the reaction, R is the ideal gas constant, and T is the temperature in degrees kelvin.

Thus the effect on the rate constants of changing the temperature from T_1 to T_2 is given by:

$$\log_{10} \frac{k_2}{k_1} = \frac{E_a}{R} \left(\frac{1}{T_1} - \frac{1}{T_2} \right)$$ (4.12)

E_a is the activation energy of the reaction, R is the ideal gas constant, T_1 is one temperature, T_2 is the other temperature in degrees kelvin, and k_1 and k_2 are the respective rate constants.

A rough estimate of the temperature effect for most reactions is that the reaction rate doubles for every $10°C$ rise in temperature. Thus the temperatures of kinetic assays are important and are normally carefully specified and controlled.

COMPARISON OF KINETIC AND EQUILIBRIUM ASSAYS

A comparison of the primary features of kinetic and equilibrium assays is shown in Table 4.6. Both systems have their advantages, and each have their advantages, and both have their disadvantages. Fundamentally either will be acceptable in most situations; however, there are assays, situations, and matrices that dictate the choice of one over the other.

TABLE 4.6.   A Comparison of the attributes of kinetic and equilibrium assays

Attribute	Kinetic Assays	Equilibrium Assays
Signal measured	Change in Absorbance with time	Absorbance
Effect of Time of Measurement	Critical	Sufficient time must be allowed to reach equilibrium
Effect of Temperature of Reaction	Important to control	Will affect time to equilibrium but usually little effect on equilibrium concentration
Key Equations	Pseudo first order kinetics, Arrhenius Equation	Law of Mass Action
Potential for Interferences	There are many potential interferences for kinetic assays, however kinetic assays can sometimes be used to avoid the slower side reactions that occur with equilibrium assays.	Lower, however the selectivity that can be obtained with kinetic assays is not normally present in equilibrium assays
Time of Reaction	Shorter	Longer
Importance of Stable Product	Less important	More important
Blanks	The Reaction Solution is often its own optical blank.	Blanks Needed
Sensitivity	Lower	Higher
Instrument Cost	Higher	Lower
Use molar absorbitivity in assays	Usually	Infrequent

SELECTION OF A COLORIMETRIC ASSAY METHOD

There are a large number of colorimetric reactions for the determination of analytes in biological systems. Frequently these assays have significantly different linear ranges. Assays that work well in one matrix are frequently inappropriate for the analysis of the same analyte in another matrix. Thus in choosing the appropriate assay method, the analyst should consider the chemistry of the analyte, the expected level of the analyte in the sample, the chemistry of the assay reaction, and the chemistry of the matrix wherein the analyte is found. The final choice should always be validated by test analysis of samples of known concentration in matrices similar or identical to the matrices of the unknown samples. Given the large number of compounds in biological samples that have similar chemistries, it should not be surprising that the use of traditional colorimetric assays often yields erroneous results. The use of enzyme-mediated colorimetric reactions can give better specificity and is discussed in Chapter 6.

SUMMARY

Colorimetric reactions can be extremely useful for the analysis of biological components in fairly complex matrices. However, the analyst needs to understand the chemistry of the analyte, the detection reaction, and of the matrix components. Given the potential for nonlinearity and matrix interference, the analyst should utilize the appropriate blanks and controls in the analysis. The positive control of spiking is strongly advised at least during the method development and validation process. (See Chapter 5). For equilibrium assays, the law of mass action is the driving force in determining the amount of product formed. In kinetic assays the pseudo–first–order kinetics of the reaction and the Arrhenius equation are the driving forces in determining the rate of product formation. In equilibrium assays the product concentration is proportional to the concentration of the analyte. In kinetic assays the rate of product formation is proportional to the concentration of the analyte. There are several areas of detection reactions which require special attention including selectivity of the detection reaction for the analyte, selectivity of the detection wavelength for the analyte, analyte loss, inhibition of the detection reaction, the effect of reagent concentrations, and time-temperature combinations.

REFERENCES

1. Kinetic Assays: D. Pérez-Bendito and M. Silva. *Kinetic Methods in Analytical Chemistry*, J. Wiley & Sons, New York, 1988.
2. H. A. Mottola. *Kinetic Aspects of Analytical Chemistry*, John Wiley & Sons, New York, 1988.

5. Enzymes and the Use of Enzymes as Reagents

INTRODUCTION

Catalysts are compounds (other than the reactants or products) that act to increase the rates of chemical reactions. Catalysts participate in the reaction being catalyzed but are regenerated at the end of the reaction sequence in their original form. Catalysts catalyze both the forward and reverse reaction of any chemical reaction and thus catalysts do not alter the equilibrium of that reaction. While there are a variety of nonbiological catalysts used for analyses in biological systems, in most cases they lack the needed specificity for the analyses of compounds in biological systems. On the other hand, many of the biological catalysts called enzymes do have the needed specificity, and modern analysts frequently use enzymes as specific reagents for the determination of individual analytes and/or classes of analytes in biological systems.

Enzymes[1] have impressive catalytic activities and typically enhance the rate of the reactions they catalyze by factors of 10^6 to 10^{20}. Values of 10^6 to 10^{12} are

---

[1] Detailed information is available on many enzymes. Excellent references include: *Methods in Enzymology*, published by Academic Press, Editors-in-Chief: Sidney P. Colowick and Nathan O. Kaplan. There are more than 110 volumes in this series, covering an extensive range of topics. *The Enzymes*, 3rd edition, edited by Paul D. Boyer, is an excellent, broad series with less emphasis on methology than *Methods in Enzymology*. *Methods of Enzymatic Analysis*, 2nd and 3rd editions, Editor-in-Chief: Hans U. Bergmeyer, published by Verlag Chemie. provides in-depth discussions of techniques of analysis that use enzymes or assay enzymes.

most common. Most chemical reactions in biological systems are catalyzed by enzymes.[2] Enzymes are composed of linear chains of amino acids called *proteins.* Different enzymes have different sequences of amino acids. The molecular weights of enzymes range from 13,700 daltons[3] to several hundred thousand daltons. The catalytic activity of enzymes depends upon their primary structure (the sequence of amino acids) as well as upon their three-dimensional structure.

Enzymes frequently have significant selectivity for given substrates. Many enzymes can readily distinguish between different constitutional isomers[4] and different stereoisomers. For example, the sugar oxidases have high substrate specificity and will typically oxidize only one type of sugar in complex sugar mixtures (see Fig. 5.1). Not only are enzyme-mediated reactions specific for the analytes; they are also quite specific in the products produced by their catalysis. For example, while traditional nonenzymatic oxidations of D-glucose will produce a variety of thermodynamically possible products that are further oxidized, the enzyme "glucose oxidase" quantitatively produces D-gluconic acid from D-glucose. Some of the enzymes involved in degradation and synthesis of biopolymers have remarkable specificity and are used extensively in the characterization of long-chain biopolymers. Examples of the uses of such enzymes are in the PCR reaction in the amplification of specific DNA primers, as restriction enzymes in genetic mapping, and different amylases for the determination of starch in the presence of cellulose. Enzyme-catalyzed reactions are among some of the most selective assays available for chemical measurements.[5]

ENZYME NOMENCLATURE

The common name for many enzymes is typically formed by adding the suffix "ase" to the compound being altered in the enzyme reaction. Thus the enzyme that catalyzes the hydrolysis of urea is called *urease,* and the enzyme that catalyzes the hydrolysis of sucrose to glucose and fructose is called *sucrase.*[6]

---

[2] Ribozymes (which are catalytic RNA molecules) are also biological catalysts.

[3] A dalton is a unit of mass 1.000 on the atomic scale. A hydrogen atom has a molecular weight of about 1 dalton.

[4] Isomers are different compounds that have the same molecular formula. Constitutional isomers are compounds in which the constituent atoms are connected in different orders. Stereoisomers differ only in the arrangement of their atoms in space. Chiral molecules are stereoisomers that are mirror images.

[5] Other highly selective techniques include chromatographic separations, electrokinetic separations, and immunochemical assays.

[6] Sucrase is also called *invertase* because its action on sucrose (common table sugar) causes the inversion of the optical rotation of a solution of sugar.

GENERAL OXIDATIONS

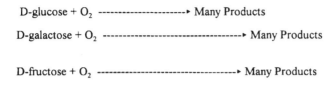

D-glucose + O_2 ---------------------→ Many Products

D-galactose + O_2 ---------------------------------→ Many Products

D-fructose + O_2 ------------------------------→ Many Products

****

ENZYMATIC OXIDATIONS

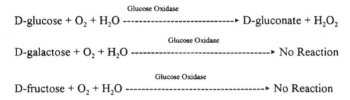

Glucose Oxidase
D-glucose + O_2 + H_2O ---------------------------→ D-gluconate + H_2O_2

Glucose Oxidase
D-galactose + O_2 + H_2O ------------------------------------→ No Reaction

Glucose Oxidase
D-fructose + O_2 + H_2O --------------------------------→ No Reaction

Fig. 5.1. Specificity of general and enzyme mediated oxidations.

Other common names are based upon the source of the enzyme. For example, *pepsin* is a proteolytic enzyme found in the stomach, *ficin* is a proteolytic enzyme found in figs, etc. Since many of the common names for enzymes lack an adequate description of the exact specificity of the chemistry of the catalysis, the use of these common names can lead to confusion and sometimes errors. The Enzyme Commission (E.C.) of the International Union of Pure and Applied Chemistry and the International Union of Biochemistry have developed a systematic naming protocol for enzymes based upon the reaction catalyzed.[7] Each enzyme is assigned a unique four-digit number. The six general classes of enzymes are listed in Table 5.1.

When one is selecting an enzyme for a given assay, it is best to use the systematic name approved by the Enzyme Commission. The use of the systematic names of the enzyme reagents should alert the analyst to potential errors and should aid in the selection of the appropriate enzyme system for a given analysis. The recent developments in the production of specialized enzymes through genetic manipulation would suggest that an analyst should also designate the source species of the enzyme, the thermostability, and perhaps whether or not the enzyme has been specifically engineered for a given analysis or matrix.

---

[7] Published by International Union of Pure and Applied Chemistry and the International Union of Biochemistry.

TABLE 5.1.   The Six Major Groups of Enzyme Activites

Group 1—oxidoreducatases	Transfer hydrogen atoms, oxygen atoms, or electrons from one substrate to another
Group 2—transferases	Transfer chemical groups between substrates
Group 3—hydrolases	Catalyze hydrolytic reactions
Group 4—lyases	Cleave substrates by reactions other than hydrolysis
Group 5—isomerases	Interconvert isomers by intramolecular rearrangements
Group 6—ligases (synthetases)	Catalyze covalent bond formation with the concomitant breakdown of a nucleoside triphosphate

ENZYMES AS REAGENTS (ENZYME-MEDIATED COLORIMETRIC REACTIONS)

Enzyme-mediated colorimetric reactions (i.e., the use of enzymes as reagents in colorimetric reactions) are attractive because of the high specificity of enzymes for given substrates, the high specificity of products produced from the analytes, and because of the impressive catalytic activity of enzymes. Enzyme-mediated colorimetric assays can use almost any detection system; absorption, fluorometric, and chemiluminescent spectrophotometry are the most commonly used.

There are thousands of different enzymes. If an analyst needs to analyze for a compound that is found in nature, there are sure to be enzymes that synthesize the compound and others that degrade it. Chemical analyses of many compounds that are not found in nature can still be achieved using enzymes. For example, one so-called "synthetic substrate" is *p*-nitrophenylphosphate, whose hydrolysis is catalyzed by alkaline phosphatase to yield phosphate and *p*-nitrophenol (see Fig. 5.2).

The use of "indicator" (or coupled) enzyme assays has become common and has greatly simplified the development of new enzyme-mediated colorimetric assays (see Fig. 5.3). Enzymes can be used to catalyze cyclic reactions and thus act as analyte concentration amplifiers. This type of use is found with both enzyme-linked immunosorbant assays (ELISA)[8] and in detection of very low

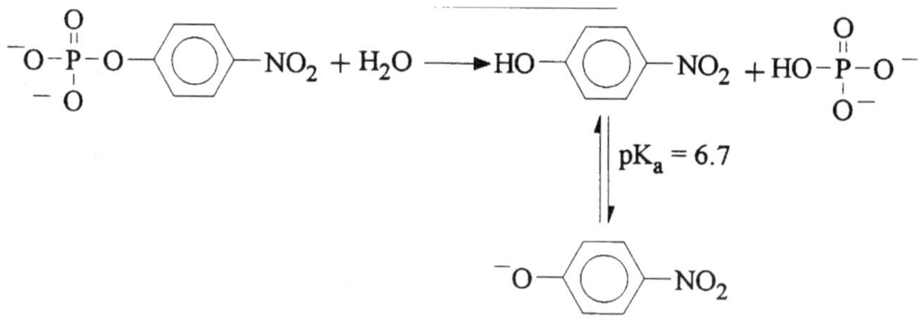

Fig. 5.2.   The hydrolysis of *p*-nitrophenylphosphate and subsequent production of *p*-nitrophenolate.

$$S_1 + S_2 \text{ --------} \rightarrow P_1 + P_2 \qquad \text{Analyte Specific Enzymatic Reaction}$$

$$P_1 + S_3 \text{ --------} \rightarrow P_3 + P_4 \qquad \text{Indicator Reaction Specific for Product } P_1$$

Fig. 5.3. Coupled reactions.

levels of analyte[9]. The use of coupled enzyme assays and cyclic reactions will be discussed later in this chapter.

Given these valuable attributes of the use of enzyme-mediated colorimetric assays, it should not be surprising that in recent years there has been an increasing use of enzymes as reagents for the assay of specific analytes in biological systems. Such use is certain to increase given the increasing availability of inexpensive sources of these enzymes developed and produced through biotechnology. There is a whole industry that develops and sells enzymes as reagents for selective analyses in biotechnology, medicine, agriculture, forensics, etc.

TYPES OF ASSAYS USING ENZYMES AS REAGENTS

The three common types of enzyme-mediated colorimetric assays are endpoint assays (quantitative conversion of analyte to product), kinetic assays (the rate of the reaction is proportional to the concentration of the analyte), and inhibition assays[10] (the percent inhibition of the rate of reaction is proportional to the concentration of the analyte).

ENDPOINT ASSAYS

In endpoint assays, the analyst typically uses an excess enzyme concentration and allows the reaction(s) to proceed until 99% (or more) of the analyte is converted to the desired product. Endpoint assays have two phases. The first is the enzymatic reaction wherein the analyte participates as a substrate, product, or a cofactor. The second phase is the detection of the product by some physical process either directly or after further chemical or enzymatic reactions. Both phases can occur concurrently. The simplest cases are those where the reaction equilibrium[11] favors the quantitative conversion ($>99\%$) of the analyte to the

---

[8] See Chapter 9.

[9] See the discussion of enzymic cycling in the section on endpoint assays in this chapter.

[10] Inhibition assays will not be discussed in this book.

[11] Remember, enzymes do not change the equilibria of reactions. They only influence the speed with which equilibrium is attained.

product and the product is directly detectable. For example, the catalysis at pH 10 of the hydrolysis of *p*-nitrophenylphosphate by alkaline phosphatase yields phosphate and *p*-nitrophenol, and then the *p*-nitrophenol is nonenzymatically rapidly deprotonated to *p*-nitrophenolate (see Fig. 5.2). The more complex cases are those in which reaction equilibrium is unfavorable or the product is not directly detectable.

SINGLE-STEP ASSAYS

When the reaction equilibrium favors the quantitative conversion of the analyte to the product and the product is directly detectable, the analyst needs to determine the amount of enzyme needed to bring the reaction to equilibrium in a reasonable time and to select the means of measuring the product concentration. While theoretically one would like always to convert 100% of the analyte to product, in practice conversions of 99.9% or 99% are acceptable.[12] Examples of the computations needed to determine the time needed for a 99% conversion are shown below. While most enzyme reactions require at least two substrates, the discussion below assumes that one of the substrates is present in great excess when enzymes are being used for reagents. If one of the substrates is not present in great excess, the computation of the time required is much more complex. The reader is referred to the discussion in Bergmeyer's *Methods of Enzymatic Analysis*.

With simple enzyme systems the kinetics of the reactions are described by the Michaelis–Menten Equation:

$$v = \frac{V_{max}[S]}{K_m + [S]} \tag{5.1}$$

where v is reaction rate or velocity, v_{max} represents the maximum rate of the reaction, [S] is the substrate concentration, and K_m is the Michaelis constant. $v_{max} = [E_{tot}]k_{cat}$.; where $[E_{tot}]$ = the total enzyme concentration and k_{cat} is the catalytic constant.

Case 1: $[S] \gg K_m$. If $[S] \gg K_m$ for the entire reaction, then the following situation holds and the time to convert 99% of the analyte to product can be calculated directly:

$$v \approx V_{max} = [E_{tot}]k_{cat} \tag{5.2}$$

---

[12] As a practical matter, given the potential for inhibition of the enzymatic process, many analysts either double or triple the amount of enzyme used or double or triple the reaction time to ensure that the colorimetric reaction has gone to equilibrium.

For example, if the k_{cat} is 20 μmol/min/μg enzyme and a 99% conversion of the analyte to product requires the conversion of 99 μmol/mL analyte to produce 99 μmol/mL product, then the

$$\text{Amount Converted} = V_{max} {}^{*}time$$

$$\text{time} = \frac{\text{Amount Converted}}{V_{max}}$$

$$\text{time} = \frac{\text{Amount Converted}}{E_{total} {}^{*}k_{cat}}$$

Therefore to convert 99 μmol/mL analyte in 1 min,

$$1 \text{ min} = \frac{(99\mu mol/mL)}{(E_{total})^{*}(20\mu mol/min/\mu g \text{ Enzyme}}$$

$$(E_{total}) = \frac{(99\mu mol/mL)}{(1 min)^{*}(20\mu mol/min/\mu g \text{ Enzyme})}$$

$$(E_{total}) = 4.95\mu g/mL$$

or since $k_{cat} = 20$ μmol/min/μg, and one unit of enzyme will produce 1 μmol/min product. Therefore, 1 μg = 20 units of enzyme. Thus you need 99 units/mL of enzyme to convert 99 μmol/mL in 1 min.

Case 2: [S] ≪ K_m. If [S] ≪ K_m for the entire reaction, then one should use:

$$v \approx \frac{V_{max}[S]}{K_m}$$

$$v = [S]\left[\frac{V_{max}}{K_m}\right] \tag{5.3}$$

Note that this is a form of a first-order reaction where $v = k[A]$. $S_o =$ initial concentration and $S =$ final concentration

$$t = \frac{1}{k}\ln\left[\frac{S_0}{S}\right]$$

Since we want 99% conversion,

$$t = \frac{1}{k}\ln\left[\frac{100}{1}\right] = 4.6\left[\frac{V_{max}}{K_{max}}\right]$$

Again if the k_{cat} is 20 μmol/min/μg enzyme and the reaction time is 1 min.

$$1 \text{ min} = \frac{4.6[K_m(\text{mmol/mL})]}{k_{cat}(\mu\text{mol/min/}\mu\text{g})E_{tot}(\mu\text{g/mL})}$$

$$= \frac{4.6 \times 10^3[K_m(\mu\text{mol/mL})]}{20 \ (\mu\text{mol/min/}\mu\text{g})[E_{tot}(\mu\text{g/mL})]}$$

$$[E_{tot}(\mu\text{g/mL enz})] = 230 \, K_m(\mu\text{g/mL})$$

COUPLED REACTIONS

When the reaction equilibrium does *not* favor the quantitative conversion of the analyte to the product and/or the product is *not* directly detectable, it is still frequently possible to use enzymes as reagents by coupling the enzymatic analyte-specific reaction to an indicator reaction (see Fig. 5.3). The specific reaction is one that reacts only with the analyte to produce a product common to many reactions. Indicator reactions quantitatively convert the product of the initial reaction to one that can be detected. Coupled reactions may be either enzymatic or nonenzymatic. A common assay is the coupled enzyme assay used for the determination of glucose as shown in Fig. 5.4. In the specific reaction the enzyme *glucose oxidase* catalyzes the production of hydrogen peroxide, which is a common product of the oxidase enzymes.[13] The indicator reaction uses horseradish *peroxidase* (an inexpensive and rugged enzyme) to convert the hydrogen peroxide and a dye precursor into a colored product and water. Other coupled reactions, including the luminol reaction and the luciferase reaction with chemiluminescence detection, are examples of very sensitive indicator reactions for hydrogen peroxide or ATP (and O_2), respectively (see Fig. 5.5). Chemiluminescence detections can be quite specific and are frequently some of the most sensitive assays available for the determination of an analyte.

Glucose oxidase

D-glucose + O_2 + H_2O ---------------------------------► D-glutanate + H_2O_2

Peroxide

H_2O_2 + o-dianisidine -------------------------► $H2O$ + oxidized o-dianisidine
(Colored)

Fig. 5.4. Glucose oxidase assay.

---

[13] Not all oxidase enzyme have hydrogen peroxide as a product. Cytochrome c oxidase produces water.

Luminol Reaction

$$\text{Luminol} + H_2O_2 \xrightarrow{\text{Microperoxidase}} \text{Aminophthalic Acid} + N_2 + \text{light}$$

Luciferase Reaction

$$\text{Luciferin} + \text{ATP} + O_2 \xrightarrow{\text{Luciferase}} \text{Oxyluciferin} + \text{AMP} + \text{PPi} + CO_2 + \text{light}$$

Fig. 5.5. Two chemiluminescence reactions sometimes used as indicator reactions in coupled reactions.

The use of PMS and MTT in the alcohol dehydrogenase assay is an example of nonenzymatic coupled reactions for visible colorimetry. Alcohol dehydrogenase catalyzes an oxidation–reduction reaction in which ethanol (CH_3CH_2OH) is oxidized to acetaldehyde (CH_3CHO) (see Fig. 5.6). This reaction can be used to determine the total ethanol in a sample. The coenzyme, nicotinamide adenine dinucleotide, is also involved in this reaction. This coenzyme exists in an oxidized form (NAD^+) and a reduced form (NADH). The oxidized form of the coenzyme accepts two electrons and a proton from ethanol to produce the reduced form of the coenzyme and acetaldehyde, or vice versa. The primary enzyme reaction can be monitored directly by measuring the absorbance change at 340 nm with time, since NADH absorbs light of this wavelength while NAD^+ does not. However, it is technically difficult to make such measurements. Furthermore, the thermodynamic equilibrium constant of this reaction favors the production of ethanol and NAD^+, and the product NADH inhibits the enzyme

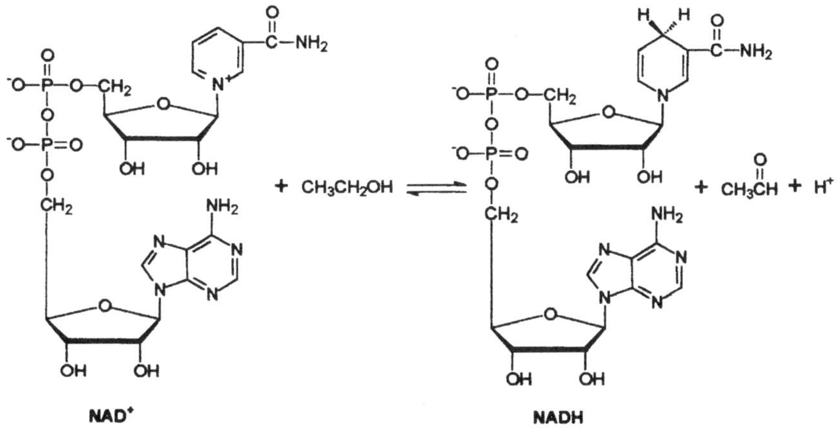

Fig. 5.6. Alchohol dehydrogenase reaction.

alcohol dehydrogenase. This is an excellent example of the type of reaction that should be monitored using coupled enzyme assays.

In the analysis of alcohol with this enzyme system, a nonenzymatic reaction couples the oxidation of NADH, formed in the primary reaction, to the reduction of a dye. The color of the reduced dye can be monitored in the visible region of the spectrum (Fig. 5.7). There are several other advantages of this method. First, not only is the concentration of NADH effectively maintained at zero, but the concentration of NAD^+ is maintained at a constant value. In addition, the molar absorptivity of the reduced dye MTT is greater than that of NADH. Thus the dye-coupled assay is more sensitive than the assay in which NADH is monitored directly. The indicator reaction uses two redox dyes, phenazine methosulfate (PMS) (Fig. 5.8) and 3-(4,5-dimethylthiazol-2-yl)-2,5-diphenultetrazolium (MTT) (Fig. 5.9). The reduction of MTT to MTT formazan, 1-(4,5-dimethylthiazol-2-yl)-3,5-diphenylformazan, is monitored in the enzyme activity assay. However, since MTT cannot be reduced directly by NADH, (Fig. 5.7) PMS is used as an intermediary to accept electrons from NADH and donate them to

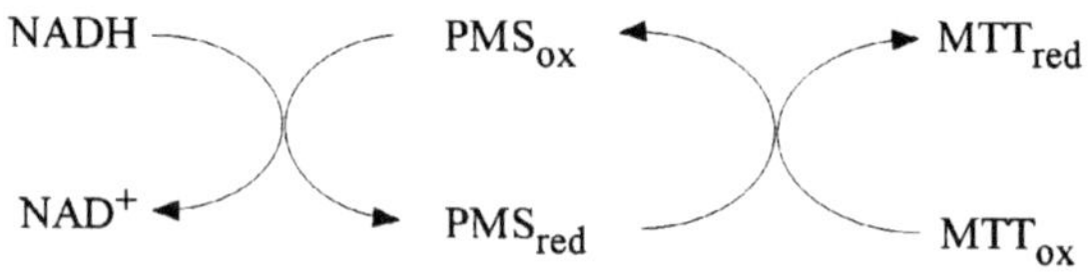

Fig. 5.7. Coupled non-enzymatic reaction for determination of NADH. The structures of PMS and MTT are shown in Figures 5.8 and 5.9.

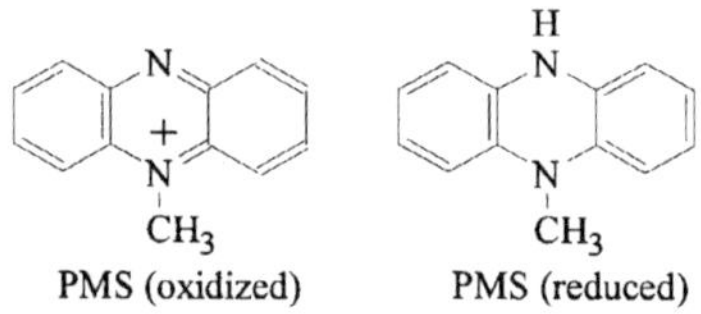

Fig. 5.8. Structure of phenazine methosulfate (PMS).

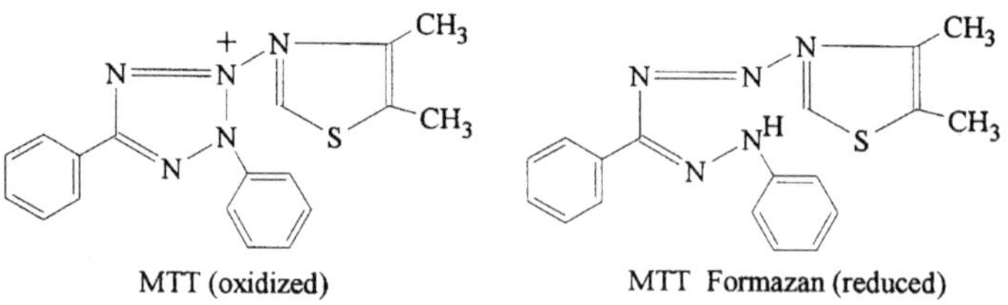

Fig. 5.9. Structure of 3-(4,5-dimethylthiazol-2-yl)-2,5-diphenyltetrazolium (MTT).

MTT. Thus the use of coupled reactions can have a variety of advantages such as producing a detectable product, converting reactions with unfavorable equilibria into coupled reactions in which the final product is favored by the equilibria of the combined reactions, maintaining of the constant levels of nonanalyte substrates for the primary reaction, and avoiding the problems that can occur when there is product inhibition of the primary reaction. The use of "indicator" (or coupled) enzyme assays has become common and has greatly simplified the development of new enzyme-mediated colorimetric assays.

CYCLIC REACTIONS

Often the concentration of the analyte is so low that even when the analyte is quantitatively converted to product, that product concentration is too low to be detected. In some cases such low levels can still be determined if cyclic assays can be used. Cyclic reactions occur wherein the analyte can be regenerated from its product. A typical system for such a reaction is shown in Figure 5.10. If the enzymes catalyzing reactions 1 and 2 are present in sufficient quantities and if the substrates S_2 and S_3 are present in excess, then the amount of P_2 and P_3 produced should be proportional to the original concentration of S_1. Cyclic assays need careful control of assay timing, and calibration curves need to be established to correlate the amount of product produced in a fixed time to the amount of analyte originally present in the sample. There are a number of cyclic assays for the determination of NADH/NAD$^+$, NADPH/NADP$^+$, and ATP/ADP.

An example of a cyclic enzymatic assay is the determination of the coenzyme NADPH by the two using glutamate dehydrogenase (E.C. 1.4.1.3) (Reaction 1) and glucose-6-phosphate dehydrogenase (E.C. 1.1.1.49) (Reaction 2), as shown in Figure 5.11. After the reactions have proceeded for a specified amount of time, the reactions are stopped and the glutamate or 6-phosphogluconate concentration is determined in a separate reaction. Cyclic reactions have been reported for the determination of pico molar[14] NADP$^+$.

$$S_1 + S_2 \ \text{---------->} \ P_1 + P_2 \qquad \text{reaction 1}$$

$$P_1 + S_3 \ \text{---------->} \ P_3 + S_1 \qquad \text{reaction 2}$$

Net reaction: $\qquad S_2 + S_3 \ \text{---------->} \ P_2 + P_3$

Fig. 5.10. A cyclic reaction with regard to S_1.

[14] 10^{-12} mol/L.

$$\text{NADPH} + \alpha\text{-ketoglutarate} + NH_4^+ \xrightarrow{\text{Glutamate dehydrogenase}} \text{glutamate} + NADP^+$$

$$NADP^+ + \text{glucose-6-phosphate} \xrightarrow{\text{Glucose-6-Phosphate Dehydrogenase}} \text{NADPH} + \text{6-phosphogluconate}$$

Fig. 5.11. A cyclic reaction for NADPH.

BIOPOLYMER DIGESTION

Endpoint reactions are frequently used when the digestion of complex polymers (or the synthesis of complex polymers) is a part of the assay. Measurement of RNA, DNA, protein, and starch is commonly done using a preliminary digestion reaction followed by the analysis of one or more of the hydrolysis products. A classic example of the use of an endpoint assay for biopolymer analysis is the determination of starch in the presence of cellulose. Starch is a mixture of polymers of glucose monomers linked α-D-$(1 \rightarrow 4)$ and occasional α-D-$(1 \rightarrow 6)$ linkages. Cellulose is a mixture of polymers of glucose monomers linked β-D-$(1 \rightarrow 4)$. Classic nonenzymatic acid hydrolysis of both starch and cellulose yields glucose. However, if these polymers are treated with the enzyme amyloglucosidase, the starch is hydrolyzed and the cellulose is not (see Fig 5.12). Amyloglucosidase catalyzes the hydrolysis of α-D-$(1 \rightarrow 4)$ and α-D-$(1 \rightarrow 6)$ glucose linkages to produce glucose but does not catalyze the hydrolysis of β-D-$(1 \rightarrow 4)$ glucosidic linkages.

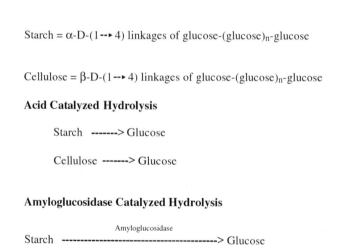

Starch = α-D-$(1 \dashrightarrow 4)$ linkages of glucose-(glucose)$_n$-glucose

Cellulose = β-D-$(1 \dashrightarrow 4)$ linkages of glucose-(glucose)$_n$-glucose

Acid Catalyzed Hydrolysis

Starch $\xrightarrow{\hspace{1cm}}$ Glucose

Cellulose $\xrightarrow{\hspace{1cm}}$ Glucose

Amyloglucosidase Catalyzed Hydrolysis

Starch $\xrightarrow{\text{Amyloglucosidase}}$ Glucose

Cellulose $\xrightarrow{\text{Amyloglucosidase}}$ No Reaction

Fig. 5.12. A starch assay.

The digestion of complex polymers is frequently a time-consuming process, and it is important that the reaction times are adequate for complete digestion. It is important that the time course of the digestion reactions be determined and that the digestions are quantitative. Assays that do not use quantitative digestions frequently have very poor recoveries and often have unacceptable precisions. In the case of the digestion of potato starch with amyloglucosidase, reaction times of 18 to 24 h are common.

KINETIC ASSAYS

The enzymatic determination of an analyte concentration can also be done by measuring the rate of the reaction and correlating it to the concentration of the analyte. What is usually done is to develop a standard curve under conditions where the rate of the reaction is proportional to the concentration of analyte. The majority of such assays are done under those conditions where the rate of the reaction is directly proportional to the concentration of the analyte. These are the pseudo-first-order reactions. There are two types of pseudo-first-order reactions: enzyme catalyzed and non-enzyme catalyzed. With enzyme-catalyzed reaction, pseudo-first-order reactions occur when $[S] \ll K_m$. In such cases the Michaelis–Menten kinetic equation simplifies to that shown in Eq. 5.3 and:

$$v = k_1[S] \tag{5.4}$$

where $k_1 = V_{max}/K_m$.

Non-enzyme-catalyzed reactions of the type shown in Figure 5.13 become pseudo-first-order if one of the reactants is present in large excess and the reaction is measured before any significant amount of product is formed (initial velocity conditions); then the normal kinetic equation describing the velocity of the reaction can be approximated by a first-order equation:

$$A + B \underset{k_{-1}}{\overset{k_1}{\rightleftarrows}} C + D$$

Fig. 5.13. A classical non-enzymatic reaction.

$$\frac{dC}{dt} = [A][B]k_1 - [C][D]k_{-1}$$

$$\frac{dC}{dt} = [A]k_1' \tag{5.5}$$

The initial velocity is measured, $[B] \gg [A]$, and $k_1' = [B]k_1$.

With the kinetic assays using the pseudo-first-order kinetics, a standard curve is developed wherein the *velocity* of the reaction is plotted versus the analyte concentration, as shown in Figure 5.14.

Coupled reactions can also be used for kinetic assays. The use of coupled reactions in kinetic assays has the same advantages as those for endpoint assays in that they can produce a detectable product, overcome the problems related to unfavorable equilibria, maintain the constant levels of nonanalyte substrates for the primary reaction, and avoid the problems that can occur when there is product inhibition of the primary reaction. Just as in endpoint reactions, both enzymatic and nonenzymatic reactions can be used for the coupling of kinetic assays. Kinetic assays can also be done under non-pseudo-first-order conditions such as zero-order kinetics, enzymic cycling, second-order kinetics conditions, etc. Inhibition of enzymic reactions can also be used to determine analyte concentrations. Non-pseudo-first-order reactions and inhibition assays can be quite complex and are beyond the scope of this book. The reader should consult a more detailed book, such as Bergmeyer's *Methods of Enzymatic Analysis.*

Kinetic assays usually take less time to do than endpoint assays and can have even better specificity. However, kinetic assays are more susceptible to matrix inhibition and do require more expensive instrumentation. Good kinetic assay methods have built in analytical quality-control procedures.

SELECTIVE SYNTHESIS OF BIOPOLYMERS

Recent advances in molecular biology and biotechnology have led to the use of selective synthesis of DNA and RNA as analytical probes for specific sequences of these important biopolymers. While such techniques are very important, they

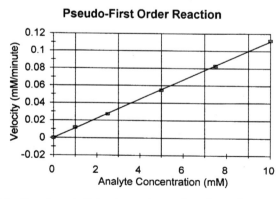

Fig. 5.14. Use of pseudo-first order reactions to determine analyte concentrations.

are outside the scope of this book, and the reader is referred to current books on laboratory techniques for molecular biology for further details.

SPECIFICITY OF ENZYMES

A critical component of determining the appropriate enzymes for use in enzyme-mediated colorimetric reactions is the specificity of the enzymes for the analyte. Obviously the enzyme(s) should catalyze the conversion of the analyte into the detectable product but should *not* catalyze the conversion of other components of the matrix into a detectable product. The first issue to resolve in the selection of the enzyme is an exact definition of the analyte. The chemical structure of the analyte needs to be sufficiently defined so that the analyst can determine if the enzyme system will catalyze the desired reactions. In most cases biological definitions such as vitamin B_6 are inadequate, since there are several chemical isomers that have the biological activity associated with vitamin B_6. Many enzyme-mediated colorimetric assays will not convert all forms of the vitamin into the detectable product.

The second issue is the exact specificity of the enzymes being used in the assay. In most cases trivial names do not provide a sufficient definition of the exact reaction that is catalyzed, and the E.C. names are needed for that exact definition. For example, while one might assume that only the sugar sucrose would be cleaved by the enzyme sucrase, this is not the case, since raffinose and stachyose are also cleaved by sucrase. The proper name for sucrase is β-fructofuranosidase (E.C. 3.2.1.26). The enzyme nomenclature manual notes that this enzyme catalyzes the hydrolysis of terminal nonreducing β-D-fructofurano-side residues in β-fructofuranosides. As shown in Figure 5.15, sucrose, raffinose, and stachyose all have β-D-fructofuranoside residues, and thus all three oligosaccharides will be cleaved by sucrase and fructose will be one product of such hydrolyses. The other products of the sucrase cleavage of sucrose, raffinose, and stachyose are different (see Fig. 5.16). Note that if the indicator reaction measures fructose, then an analyst would get positive responses from each of these oligosaccharides. On the other hand, if the indicator reaction measures the released glucose, then an analyst could measure the level of sucrose in the presence of raffinose and stachyose. In any case, if the analyst knows that compounds structurally similar to the analyte are likely to be present in the sample matrix, it is prudent to check the specificity of the enzyme assay system to ensure that either only the analyte is acted upon by the enzyme(s) catalyzing the colorimetric reaction or that a unique product is produced from the analyte.

Finally, the analyst needs to know whether or not the matrix contains compounds other than the analyte that are substrates for the enzymes used to catalyze the detection reaction. If the analyst knows that a matrix does not contain substrates for the enzyme other than the analyte, then an enzyme assay system

Sucrose is 1-α-D-glucopyranosyl-β-D-fructofuranoside;

Raffinose is α-D-galactopyranosyl(1-6)-α-D-glucopyranosyl(1-2)-β-D-fructofuranoside

Stachyose is α-D-galactopyranosyl(1-6)-α-D-galactopyranosyl(1-6)-α-D-glucopyranosyl(1-2)-β-D-Fructofuranoside

Fig. 5.15.   Some sugars yielding fructose when acted upon by sucrase.

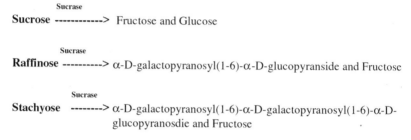

Sucrase
Sucrose ------------> Fructose and Glucose

Sucrase
Raffinose ----------> α-D-galactopyranosyl(1-6)-α-D-glucopyranside and Fructose

Sucrase
Stachyose --------> α-D-galactopyranosyl(1-6)-α-D-galactopyranosyl(1-6)-α-D-glucopyranosdie and Fructose

Fig. 5.16.   Products of sucrase actions on several sugars.

can be used for an analysis even if it lacks high specificity. For example, an assay combining sucrase and a fructose quantifying reaction can be used to determine the total sucrose in most fruits because they do not contain raffinose and stachyose. On the other hand, if soy beans are the matrix of concern, then the analyst needs to be aware that soybeans contain sucrose, raffinose, and stachyose and that the use of an asay combining sucrase and a fructose quantifying reaction for the determination of sucrose will yield high, inaccurate results. See figure 5.17.

Most Fruits:

Sucrose is the only sugar whose hydrolysis which is catalyzed by *sucrase* to yield fructose.

Soybeans:

Contain three sugars whose hydrolysis which is catalyzed by *sucrase* to yield fructose. Only one of these is sucrose

Fig. 5.17.   The effect of matrix on the analysis for sucrose.

ENZYME INACTIVATION

Some enzymes are easily denatured and inactivated. Disruption of either the primary structure or the three-dimensional structure of an enzyme (enzyme denaturation) will usually destroy the catalytic activity of the enzyme. Denaturation can be caused by a wide variety of factors, including heat, changes in pH, phase interfaces, changes in polarity of the solvents, freeze-thawing, glass surfaces, detergents, and dilute enzymes are more susceptible to denaturation than concentrated solutions; however, the addition of an excess of a protein[15] such a bovine serum albumin (BSA) can provide some protection against denaturation. Enzyme activity also can be irreversibly lost due to heavy metal poisoning, the actions of proteolytic[16] enzymes, the so-called "suicide substrates", oxidation, and a variety of other chemical reactions. Enzyme catalytic activity can also be lost due to the reversible binding of numerous compounds, including those that mimic the enzyme's substrates and bind to the catalytic center of the enzyme, blocking the availability of that site to the substrate of the enzyme reaction.

VALIDATION OF THE ACCURACY OF ENZYME-MEDIATED COLORIMETRIC REACTIONS

Since there are so many occurrences that could result in an enzyme-mediated colorimetric reaction giving inaccurate data, including enzyme inactivation and cases where the matrix contains some nonanalyte compound that gives a false positive signal, enzyme-mediated analyses need to be carefully validated. The analyst needs to demonstrate that the enzyme preparations have the desired activity for the assay. Proper sample blanks, enzyme blanks, and positive control samples are frequently needed in the analysis of biological materials to ensure that the results of the enzyme-mediated colorimetric reactions are accurate.

Positive Controls

An analyst uses positive controls to demonstrate that the enzymes are active and have the required reactivity in the analysis being preformed. A standard positive control is to develop a standard curve with each assay set. Note that the development of a standard curve with each analysis set can serve two purposes: (1) the creation of the analysis parameters needed to compute and evaluate the concentrations in the samples and (2) as a positive control for the analysis. The comparison of the figures of merit for the standards in the current analysis set with the figures of merit for the standards in previous analysis sets using the same

---

[15] Obviously such an added protein should not interfere with the enzyme reaction being studied.

[16] Proteolytic enzymes catalyze the hydrolysis of peptide bonds.

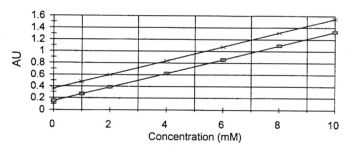

Fig. 5.18. Plot of method of standard additions data where there is no interference from the matrix. The boxes indicate the results for the standard curve and the "x"s indicate the results from the method of standard additions.

assay is the most commonly used positive control. Good analysts routinely use this type of positive control with each assay set. Obviously the desired result is that the figures of merit of the current analysis set are the same as those figures of merit determined in previous analysis sets.

Method of Standard Additions

The method of standard additions[17] is used to test the assumption that the enzymes have the same activity *in the presence of the analyte matrix* that they do in the absence of the matrix. The method of standard additions is done by analyzing a series of concentrations of the standards in the presence of the analyte matrix; that is, the method of standard additions consists of developing a standard curve in the presence and absence of the sample matrix. The product signal is then plotted versus the amount of added standard, and the slopes of the response curves in the presence and in the absence of the matrix are computed. If the slopes are the same, then this is taken as evidence that there is no interference with the enzyme assay caused by the matrix. If there is no interference, the determination of the unknown concentrations is done in the standard manner as in the standard detection reactions. However, if the slopes are not the same, this is taken as evidence that there is interference with the enzyme assay caused by the matrix. Figure 5.18 shows a standard additions plot, which shows no interference, and Figure 5.19 shows a standard additions plot where there is interference. In addition to visual inspection, various statistical tests such as overlap of the 95% confidence limits (see Appendix (2) and Student's T test are frequently used to compare the figures of merit.

The reaction tables for the method of standard addition analyses are different from the more normal reaction tables. A sample reaction table for the method of standard additions evaluation of whether or not the matrix causes interferences is

---

[17] The method of standard additions is sometimes called "spiking".

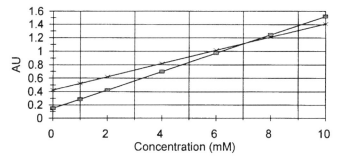

Fig. 5.19.   Plot of method of standard additions data where there is interference from the matrix. The boxes indicate the results for the standard curve and the "x"s indicate the results from the method of standard additions.

shown in Table 5.2 for an assay reaction that uses 1 mL of sample and 4 mL of reagent. The solutions for the determination of the standard curve for this reaction table were prepared as shown in Table 5.3. There are several items to note in this reaction table. First, there are two extra columns in the reaction table: one for the addition of the sample in its matrix and one for the addition of diluent. Second, note that the volume of added standard plus the volume of added diluent plus the volume of added sample in its matrix equals the total volume of the sample used in the assay. Furthermore, the volume of the added standards is half the total assay sample volume. Next it should be noted that the concentration of the standards in this reaction table is half that of the standards as prepared in the dilution table, since these standards are diluted by an equal volume of either diluent or the sample in its matrix. This reaction table uses sample 1 as a blank, since it contains neither the sample in its matrix nor any concentration of the standard, but the reagent has been diluted by the same amount as the other samples in the reaction table. Finally, it should be noted that sample 5 is the sample whose absorbance

TABLE 5.2.   A Reaction Table for Method of Standard Additions

Sample No.	Added Known Analyte (mg/mL)	Standard Added	mL of Standard of Known Concentration	Sample in its Matrix (mL)	Diluent (mL)	Reagent (mL)	Blank No.
1	0	A	0.5	0.0	0.5	4.0	1
2	1.0	B	0.5	0.0	0.5	4.0	1
3	2.0	C	0.5	0.0	0.5	4.0	1
4	4.0	D	0.5	0.0	0.5	4.0	1
5	0.0	A	0.5	0.5	0.0	4.0	1
6	1.0	B	0.5	0.5	0.0	4.0	1
7	2.0	C	0.5	0.5	0.0	4.0	1
8	4.0	D	0.5	0.5	0.0	4.0	1

TABLE 5.3. A Dilution Table for Method of Standard Additions[a]

Sample No.	Known Analyte Concentration (mg/mL)	Stock Standard (mL)	Diluent (mL)	Total Volume (mL)
A	0	0.0	2.0	2.0
B	2.0	0.4	1.6	2.0
C	4.0	0.8	1.2	2.0
D	8.0	1.6	0.4	2.0

[a]Stock analyte concentration = 10 mg/mL.

would be used to compute the original concentration of the analyte in the sample in its matrix.

There is significant disagreement as to what should be done if there is evidence of matrix interference. That discussion is best left to a more advanced discussion of analytical techniques. Suffice it to say that it is usually best to find another assay for the analyte. Use of the method of standard additions in routine analysis can significantly increase the sample load and thus the cost of the analyses and the time for these analyses. For these reasons analysts are frequently reluctant to use the method of standard additions on a routine basis. An acceptable compromise is frequently found by using the method of standard additions during the method testing and validation stage, but not after the assay method has been validated for a given matrix.

The assumption that whatever interferences there are for the assay of the analyte in the unknown sample matrix, there will be the same interferences for the assay of the added standard in the same matrix is a dangerous one, for there are examples that such assumptions are incorrect. The wise analyst presents the results of such computations with a qualification that while these were the best data available from the analyses, the data should be viewed as still potentially flawed.

ANALYTE LOSS

There are occasions in which the analytes are consumed by side reactions. If the analytes are consumed or otherwise not available for reaction, the results will be low. If the losses of analyte are proportional to analyte concentrations, spiking and other recovery studies may permit detection of the losses. If the losses are fixed and at a low level, it is rare that proper correction factors can be established. Usually the best that can be done is for the analyst to know the chemical composition of the sample and take care to use assays appropriate for the matrix and analyte.

6. Measurement of Enzyme Activities

REASONS FOR MEASUREMENT OF ENZYME ACTIVITIES

Enzymes are used in a wide variety of chemical activities, including the measurement of specific enzyme activities as clinical diagnostic tests and their use as analytical reagents (e.g., measurement of blood or urine glucose concentration and their use as indicator molecules in ELISA systems); in experimental and commercial syntheses of various organic compounds (e.g., production of fructose from glucose); the specific degradation of biological polymers (e.g., the production of glucose from starch; the very specific cleavage of DNAs in gene mapping; the degradation of low-molecular-weight compounds (e.g., the removal of glucose from egg whites); the synthesis of specific biological polymers (e.g., the synthesis of DNA in the PCR reactions); and so forth. Given the importance of these uses, it should be obvious that there is a significant interest in the measurement of enzyme activities.

At first glance it might seem that assays that measure the total protein in the sample or immunoassays that measure the concentrations of specific proteins in a sample might be used to measure the enzyme concentrations in a sample. However, such measurements have the potential of providing inaccurate information on the enzymatic activity of the sample. Enzyme preparations are rarely pure solutions of single proteins; almost always there are multiple proteins in such mixtures. Either the enzyme preparation is rather crude and the desired activity has not been purified away from the other source proteins, or the enzyme activities have been protected by excess amounts of some nonenzymatic proteins such as bovine serum albumin. In all such cases, measurements of total protein content would lead to highly inaccurate values of the enzyme concentration. Even when the only protein in the preparation is the enzyme of interest, the enzyme activity may have been reduced due to the presence of enzyme inhibitors and/or the denaturation of the enzyme protein. Again measurements of total protein would lead to inaccurate results. Even the highly specific protein immunoassays usually cannot distinguish between active and inactive enzyme preparations. In many such cases, the total protein content might be unchanged, but the enzymatic activity could be drastically reduced. The better choice of assays is to choose those assays that measure the specific catalytic activity of the enzyme.

CLASSICAL CHEMICAL KINETICS

It is useful to discuss the theoretical components of kinetic measurements as an introduction to the measurements of enzyme activities. Classical chemistry kinetic theory attempts to answer the question. "What determines the rate of formation or destruction of any given compound?" Studies over the years have shown the following:

1. For a reaction such as that shown in Figure 6.1, the rate of formation of one product at any time (dP/dt) is proportional to the concentration of the reactant(s) (S) and the rate constant ($k_{forward}$):

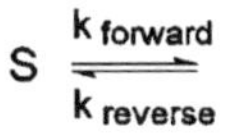

$$S \; \underset{k \text{ reverse}}{\overset{k \text{ forward}}{\rightleftharpoons}} \; P$$

Fig. 6.1.   Classical kinetics.

$$\frac{dP}{dt} = [S]k_{forward} \tag{6.1}$$

2. The rate of destruction of any product at any time ($-dP/dt$) is proportional to the concentration of the reactant(s) (P) and the rate constant ($k_{reverse}$):

$$-\frac{dP}{dt} = [P]k_{reverse} \tag{6.2}$$

3. Thus the *net rate* of formation of one compound (dP/dt) is proportional to the rate of its formation minus the rate of its destruction:

$$\text{net } \frac{dP}{dt} = [S]k_{forward} - [P]k_{reverse} \tag{6.3}$$

At equilibrium, products are formed from the reactants at the same rate that the reactants are formed from the product(s):

$$[S]k_{forward} = [P]k_{reverse} \tag{6.4}$$

Thus at equilibrium concentration of the substrates and the products do not change with time.

ENZYME KINETICS

With classical chemical kinetics, a reaction that converts a substrate (S) to a product (P) with a reaction rate (velocity) that is directly proportional to the concentration of the substrate[1] is called a *first-order reaction*. With classical first-order reactions, measuring the reaction velocity at different concentrations of substrate and plotting the velocity as a function of substrate concentration will result in a straight line (Fig. 6.2). However, when the reaction is catalyzed by an enzyme, plotting the velocity as a function of substrate concentration at constant enzyme concentration[2] results in a hyperbolic plot (see Fig. 6.2).

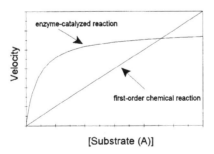

Fig. 6.2 Classical first order chemical kinetics and enzyme kinetics. Dependence of velocity on substrate concentration.

[1] While chemists refer to reactants of a reaction, biochemists refer to substrates for enzyme-catalyzed reactions. The terms substrate(s) and reactant(s) are interchangeable.

[2] In the cases shown, the concentration of enzyme is much lower than that of the substrate ([E] $\ll$ [S]).

$$E + S \underset{k_2}{\overset{k_1}{\rightleftharpoons}} ES \underset{k_4}{\overset{k_3}{\rightleftharpoons}} E + P$$

Fig. 6.3. A simple enzyme catalyzed reaction: The rate of formation of P is dependent upon the [ES] and k_3. E is the enzyme concentration, S is the substrate concentration, ES is the enzyme-substrate complex concentration, P is the product concentration, and k_1, k_2, k_3 and k_4 are the rate constants.

The hyperbolic relationship between reaction velocity and substrate concentration for the enzyme-catalyzed reaction can be understood by hypothesizing the formation of an enzyme–substrate complex (ES) as a necessary requirement for enzymatic catalysis (Fig. 6.3). When the substrate concentration is increased to level such that all the enzyme is involved in the ES complex, increasing the substrate concentration further does not increase the reaction rate. This condition is called *substrate saturation*.

The relationship between reaction velocity (v) and substrate concentration ([S]) for an enzyme that uses a single substrate is defined by the Michaelis–Menten equation[3]

$$V_0 = \frac{V_{max}[S]}{K_m + [S]} \tag{6.5}$$

V_{max} is the maximum initial velocity for any given enzyme concentration, v_0 is the initial velocity, [S] is the substrate concentration, and K_m is the Michaelis constant.

The Michaelis–Menten equation is derived for the reaction mechanism shown in Figure 6.3, with the assumption that the concentration of the product is zero and thus the rate of formation of the substrate from the product is zero. This velocity is called the initial velocity (v_0). Under initial velocity conditions, the rate of the reaction is independent of the time of the reaction (see Fig. 6.4).

The assumptions used to derive the Michaelis–Menten equation are:

1. The concentration of the enzyme is much lower than the concentration of the substrate.
2. The rate of formation of the enzyme–substrate complex (ES) is equal to the rate of its breakdown (steady-state assumption).
3. The concentration of the product is zero (initial velocity conditions).

[3] The Michaelis–Menten equation describes the simplest kinetics for enzyme-catalyzed reactions. There are many cases of enzyme-catalyzed reactions with much more complex kinetic behavior. All enzyme kinetic systems require the postulate of at least one enzyme–substrate complex.

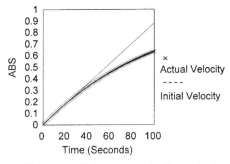

Fig. 6.4. Initial velocity (v_0) as compared with actual velocity.

4. The total concentration of the enzyme [E_{total}] is equal to the sum of the free-enzyme concentration [E] and that of the enzyme-substrate complex [ES]. (Conservation equation).

The important terms in the Michaelis–Menten equation are V_{max} and K_m. V_{max} is the maximum velocity for the enzyme-catalyzed reaction and is equal to [E_{total}]k_3. k_3 is also referred to as $k_{cat}(k_{catalysis})$ (see below). K_m, the Michaelis constant, is a kinetic constant. K_m is operationally defined as the concentration of the substrate that results in a reaction velocity equal to one-half the maximum velocity. A discussion of the implications of the theoretical components of K_m is outside the scope of this book. The reader is referred to a good biochemistry textbook.

V_{max} is a function of the enzyme concentration and the catalytic efficiency (k_{cat}) of the enzyme:

$$V_{max} = k_{cat}[E_{total}] \qquad (6.6)$$

k_{cat} is sometimes called the *turnover number* of the enzyme. The Michaelis–Menten equation (Eq. 6.5) predicts that with excess substrate (i.e., [S] $>>$ K_m), the initial reaction velocity (v_0) should equal V_{max}. If there is not excess substrate, then the Michaelis–Menten equation predicts that the initial velocity should be a hyperbolic function of substrate concentration [S]. These predictions are consistent with the experimental observations of many enzyme systems (see Fig. 6.2). Note that the net result of these discussions is that as long as the substrate concentrations are in excess, for every enzyme concentration there is a limit to the reaction velocity, which is called V_{max}. Since the initial velocity v_0 is a function of the total enzyme concentration, at different enzyme concentrations the initial rates of formation of the product will differ (see Fig. 6.5). When the initial

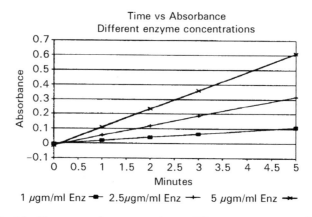

Fig. 6.5. Time course of enzyme reactions at different enzyme concentrations.

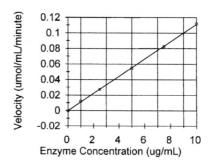

Fig. 6.6. Effect of enzyme concentration on initial velocity. The slope of this plot is the k_{cat} or turnover number.

velocities of product production are plotted versus the enzyme concentration, one should obtain a straight line, as shown in Figure 6.6.

Enzyme activity is expressed in moles of substrate lost or of product gained per unit time. (They should be the same!) Thus the amount of enzyme activity will be expressed as the change in the amount of the substrate per minute. The standard unit of enzyme activity is the amount of the enzyme that will catalyze the reaction of one μmol of substrate per minute. The *specific activity* is the amount of enzyme activity per weight of enzyme preparation and is usually expressed as μmoles/min/mg enzyme. The *turnover number* (k_{cat}) is the amount of enzyme activity per mole of enzyme and is expressed as μmoles/min/μmol enzyme.

MEASUREMENTS OF ENZYME ACTIVITIES

Enzyme concentrations are determined by analyzing the catalytic activity of the enzyme preparation.[4] Practically, enzyme activity assays are done by mixing the substrate with the buffer and other components of the assay mixture and then by adding the enzyme and immediately starting a timer. The amount of product produced is measured over time, and then the amount of product produced is plotted versus time. Given the requirement for initial velocity (v_0) conditions, only the slope of that initial straight-line portion of the plot is measured and that slope is v_0 (see Fig. 6.4).

Operationally there are two types of assays: the continuous enzymatic assay and the discontinuous assay. In a continuous assay one monitors the product formation or substrate loss as a function of time. There is no need to stop the reaction to determine the product concentration. In a discontinuous assay aliquots are removed from the reaction vessel, and the reaction is stopped prior to the measurement of the product concentration. While various measurement systems may be used to determine enzyme activity, this discussion will be based upon the spectrophotometric measurements of product formation. The theoretical concepts and constructs are the same for any mode of measurement of product formation.

CONTINUOUS ENZYME ASSAYS

A continuous enzyme assay may be used if the substrate and final product have markedly different absorption spectra in the assay media. For example, while *p*-nitrophenylphosphate has almost no absorbance at 400 nm, its hydrolysis catalyzed by alkaline phosphatase yields *p*-nitrophenol, which rapidly dissociates at the alkaline pH of the enzymatic reaction to produce *p*-nitrophenolate, which has a significant absorption at 400 nm. The reactions are shown in Figure 6.7. Thus a continuous spectrophotometric assay can be used to determine the activity of alkaline phosphatase when the substrate is *p*-nitrophenylphosphate.

DISCONTINUOUS ENZYME ASSAYS

There are many cases where the product of an enzyme catalysis is not directly detectable by spectrophotometric analysis. Such products can frequently be

---

[4] An in-depth discussion of the determination of enzyme K_ms, V_{max}s, and K_is is outside the purview of this book, and the reader is referred to a good biochemistry text.

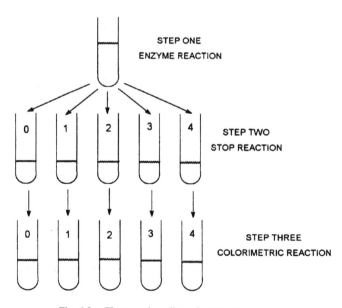

Fig. 6.7. The catalysis of the hydrolysis of *p*-nitrophenylphosphate by alkaline phosphatase. The dissociation of the *p*-nitrophenol to *p*-nitrophenolate is a non-enzymatic reaction. Note that there are two products of this reaction (*p*-nitrophenolate and phosphate) and that the kinetics of the reaction could be determined by following the change in concentration of either product.

converted into detectable products by the use of a colorimetric reaction. Such enzyme systems can be measured using discontinuous assays. In discontinuous assays aliquots are removed from the reaction mixture at a series of times after the addition of the enzyme and treated to stop the enzymatic reaction. The colored product is developed via a colorimetric reaction. Such a series of steps is shown in Figure 6.8.

Fig. 6.8.   The steps in a discontinuous enzyme assay.

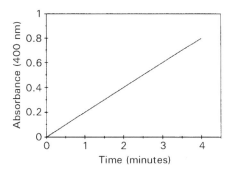

Fig. 6.9. The initial velocity of an alkaline phosphatase catalyzed hydrolysis of *p*-nitrophenyl-phosphate.

CALCULATIONS FOR CONTINUOUS ENZYME ASSAYS

A typical initial output for continuous assays is shown in Figure 6.9. In this case the plot shows the initial production of *p*-nitrophenolate with time for the action of alkaline phosphatase on *p*-nitrophenylphosphate. The slope of the line is the initial reaction velocity, whose units are Δ (absorbance) per minute ($\Delta A/$min). In this example, the velocity is 0.20 AU/min. These velocity units can be converted to a change in *p*-nitrophenolate concentration, since the molar absorptivity for *p*-nitrophenolate at 400 nm is known ($\varepsilon = 18.3$ mM^{-1} cm^{-1}). For the measurements made here the cuvette light path length was 1 cm, and the enzyme concentration was 10 μg/mL.

CALCULATIONS

A series of steps is required to compute the velocity of an enzyme-catalyzed reaction and then the specific activity of the enzyme preparation.

1. The first step is to convert the change in absorbance per minute ($\Delta A/$min) to the change in concentration per minute. This conversion involves dividing $\Delta A/$min by the molar absorptivity of *p*-nitrophenolate:

$$\frac{\Delta A}{(\text{min})(\text{cm})} \div \varepsilon(\text{mM}^{-1}\,\text{cm}^{-1}) = \frac{\Delta\,\text{mM}}{\text{min}}$$

$$\frac{0.20}{\text{min}} \div 18.3\,\text{mM}^{-1} = \frac{0.011\,\text{mM}}{\text{min}} \tag{6.7}$$

Note that the path length units of cm and the cm units of molar absorptivity cancel.

2. The second step is to convert the per liter concentration to a per mL concentration. Remember that 1 mM concentration is exactly the same as a concentration of 1 μmol/mL; thus velocity can be converted:

$$\frac{0.011\,\text{mM}}{\text{min}} = \frac{\dfrac{0.011\,\mu\text{mol}}{\text{mL}}}{\text{min}} \tag{6.8}$$

3. The third step is to compute the specific activity (i.e., μmol/min/mg enzyme) from the velocity and the enzyme concentration. There are two scenarios. In the first, only one enzyme concentration was used to determine the reaction velocity; in the second, a series of enzyme concentrations was used and the reaction velocities were determined for each enzyme concentration.

a. Single Enzyme Concentration

When a single enzyme concentration is used to determine the reaction velocity, then the specific activity equals the velocity divided by the enzyme concentration. In the example given above, 10 μg enzyme/mL (0.01 mg enzyme/mL) was used as the enzyme concentration. Thus the specific activity would be computed as:

$$\frac{\dfrac{0.011\,\mu\text{mol}}{\text{mL}}\dfrac{1}{\text{min}}}{\dfrac{0.010\,\text{mg enzyme}}{\text{mL}}} = \frac{\dfrac{0.0011\,\mu\text{mol}}{\text{min}}}{\text{mg enzyme}} = \frac{\dfrac{1.1\,\text{nmol}}{\text{min}}}{\text{mg enzyme}} \tag{6.9}$$

b. Multiple Enzyme Concentrations

When reaction velocities are measured for reactions having different enzyme concentrations, the specific activity is determined by plotting the velocities versus the enzyme concentrations and determining the slope of the plot. The slope is the specific activity.

For example, if a series of enzyme concentrations catalyzed a series of reactions, Table 6.1 might be the result. Plotting these data gives the plot shown in Figure 6.10. The slope of this plot of velocity versus enzyme concentration is the specific activity of the enzyme. The slope for these analyses is 0.0111 ± 0.0001 μmol/min/μg enzyme.

TABLE 6.1. Hypothetical Set of Velocities for Given Enzyme Concentrations

Enzyme Concentration (μg/mL)	Velocity (μmol/mL/min)
0.0	0
1.0	0.012
2.5	0.027
5.0	0.054
7.5	0.083
10.0	0.112

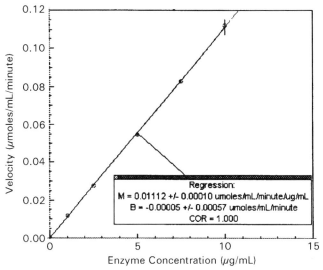

Fig. 6.10. Plot of velocity versus enzyme concentration. The slope of the plot is the specific activity of the enzyme preparation.

CALCULATIONS FOR DISCONTINUOUS ENZYME ASSAYS

The steps in the calculations for discontinuous enzyme assays are more complex than those of continuous assays, but the reader should note significant similarities.

1. The first step is to make a standard curve for the colorimetric assay and then to compute the R^2, sensitivity, LOD, and linear range for colorimetric standard curve.
2. The second step is to plot absorbance from colorimetric assays of enzyme assay aliquots versus time for each enzyme concentration.

3. The third step is to compute the slope for each plot of each enzyme concentration. This is the crude velocity, and it will have units of absorbance per time.

4. The fourth step is to divide the slopes obtained in step 3 by the sensitivity of the colorimetric standard curve (Step 1) to get the initial velocity of the analysis in the appropriate units (i.e., concentration per time):

$$\frac{\text{slope (AU/min)}}{\text{sensitivity of standard curve (AU/analyte concentration)}} = \frac{\text{analyte concentration}}{\text{min}}$$

$$(6.10)$$

5. The fifth step is to construct a table containing the enzyme concentrations and the computed velocities in concentration/time (mmoles/mL/min) or (μmol/mL/min) or (nmol/mL/min).

6. The sixth and final step is to plot the velocity versus enzyme concentration and compute the slope of this plot, which is the specific activity.

Example: You are assaying alkaline phosphatase using discontinuous assays of alkaline phosphatase activity using *p*-nitrophenylphosphate as a substrate. The velocities of the reactions are determined by using a discontinuous analysis of the released phosphate ions. The assay involves the interaction of the phosphomolybdate complex with malachite green to form an intensely colored complex. The instructions are as follows:

Standard Curve

Prepare a series of tubes, each containing a final volume of 0.4 mL with increasing amounts of inorganic phosphate (0, 4, 8, 12, 16 and 20 nmol per 0.4 mL). Add 2.6 mL of malachite reagent to each tube and mix. Wait 10 min and measure the absorbance of each sample at 645 nm. Plot a standard curve and determine the linear range of the assay.

Discontinuous Assays of AP Activity

Using 0, 1.5, 2.5, and 5 μg AP per mL of reaction and a 3 mL reaction mixture, determine the rate of phosphate formation in your assay system. Before starting these assays, appropriately label a second set of tubes that will be used for chromophore formation. Add 2.6 mL of malachite reagent to each chromophore tube. In each enzyme assay, all the reaction components *except* the enzyme should be combined in a test tube. The enzyme should be added and the tube mixed immediately. Start your timing for the enzyme tube now. As rapidly as possible, remove 0.4 mL of the reaction mixture and add it to the malachite reagent in a tube marked "0". This tube will be the zero-time sample. Start your timing for the color reaction when this 0.4 mL sample is added to the reagent.

TABLE 6.2. Data for Phosphate (P$_i$) Standard Curve

P$_i$ conc. (nmol/0.4 ml)	abs. 645 nm	P$_i$ conc. (nmol/mL)
0.0	0.000	0
4.0	0.090	10
8.0	0.215	20
12.0	0.349	30
16.0	0.469	40
20.0	0.622	50

Remove 0.4 mL aliquots of the enzyme solutions at 1, 2, 3, and 5 min after the zero-time sample was taken and add them to malachite reagent containing tubes labeled 1, 2, 3, and 5 min.[5] Thus you will have four enzyme sample reaction tubes and for each enzyme sample you will have 5 tubes to assay for the content of phosphate.

Measure the color development at 645 nm of each tube 10 min after the sample was mixed with the color reagent.

The absorbance values for the different standard phosphate concentrations are shown in Table 6.2. Note that the phosphate concentration has been converted from nmol/0.4 mL to nmol/mL. The standard curve of these data is shown in Figure 6.11.

The slope of the standard curve is thus 0.0125 AU/nmol/mL of phosphate. The absorbance data from the colorimetric determination of the phosphate in each time–enzyme concentration sample are shown in Table 6.3, along with the conversion of the data from absorbance units/min to nmol/mL/min. The data in Table 6.3 are then plotted as shown in Figure 6.12, and then regression analyses are used to determine the slopes of each of the individual plots for each enzyme concentration. These slopes are the velocities for each enzyme concentration. These velocities are then plotted versus the enzyme concentration (see Fig. 6.13). The slope of this plot is the specific activity of the enzyme as determined by the discontinuous enzyme assay.

ENZYME ASSAY REACTION CONDITIONS

The proper selection and use of specific reaction conditions are usually critical in the development and utilization of an assay for enzyme activity. Frequently components such as pH, temperature, buffer selection, ionic strength, cofactor

---

[5] Note that the addition of the reaction aliquot to the malachite reagent has two functions: one to stop the enzyme-catalyzed reaction and two to produce a colored product from the colorless product of the enzyme catalysis (phosphate).

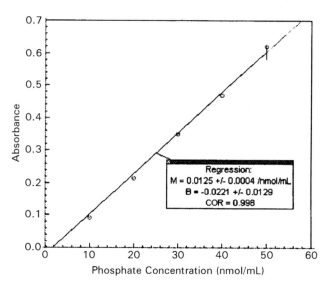

Fig. 6.11. Phosphate standard curve.

concentrations, and substrate concentrations are critical and need to be carefully selected and rigorously followed. The selection of the proper reaction conditions for the assay of enzymatic activities is a complex topic by itself, and the reader is referred to the numerous treatises on the assay of enzymes. Descriptions of many enzymatic assay systems can now be found on the World Wide Web at the sites of those who produce or sell enzyme preparations as well as other sites. Analysts are well advised not to arbitrarily change published reaction conditions. Furthermore, most assays of enzyme activity have been developed for specific matrices, and an

TABLE 6.3.   Time and Absorbance Data at Different Enzyme Concentrations

Time (min)	Concentration enzyme (μg/mL)			
	0	1.5	2.5	5
0	0	0.01	0.01	0.02
1	0	0.076	0.099	0.153
2	0	0.105	0.186	0.317
3	0	0.17	0.29	0.503
5	0	0.312	0.488	0.853
Least-squares slope (AU/min)	0.00	0.0591	0.0959	0.169
Divide by slope New units of (nmol/mL/min) standard curve (0.0125 AU/nmol/mL)	0.00	4.73	7.67	13.52

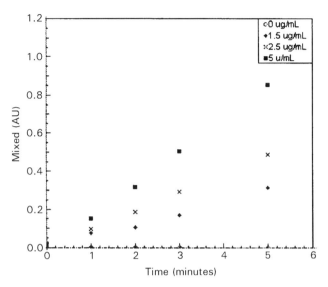

Fig. 6.12. Time courses of the alkaline phosphatase catalyzed reactions for four different enzyme concentrations.

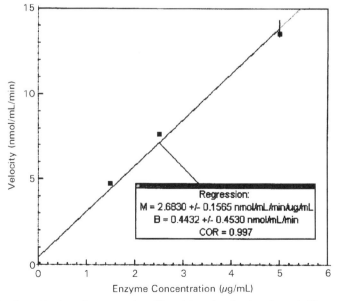

Fig. 6.13. Determination of an enzyme specific activity using the data shown in Figure 6.12.

analyst should confirm that such assays are still valid when utilized with matrices other than those used in the development of the assay.

COUPLED REACTIONS

If the reaction equilibrium does not favor the quantitative conversion of the analyte to the product and/or the product is difficult to detect (either directly or indirectly), enzyme assays may be done through the use of coupled enzyme assays (see Chapter 5). Coupled enzyme assays are constructed by coupling the enzymatic analyte specific reaction to an indicator reaction. The specific reaction is one that reacts only with the analyte to produce a product common to many reactions. The indicator reactions quantitatively convert the common product of the initial reaction to a final product that can be detected. Coupled reactions may be either enzymatic or nonenzymatic and may be used in either continuous or discontinuous enzyme assays. The determination of the conditions for using coupled enzyme assays to measure enzyme activities is beyond the scope of this book, and the reader is referred to a book such as in Bergemeyer's series, *Methods of Enzymatic Analyses.*

MULTISUBSTRATE ENZYMES

While the preceding discussion dealt with single-substrate reactions, most enzymes catalyze reactions that use two or more substrates. The kinetic equations for these enzymes are more complex than those for a single-substrate enzyme. However, some common principles apply to all enzyme assays. For example, high concentrations of substrates are required to approach V_{max} for all enzymes. Thus the assay of multisubstrate enzymes requires that all the substrates be at a level where the enzyme is saturated with its substrates. Thus, even with these multisubstrate enzymes, it is possible to evaluate the amount of the enzyme in an unknown using V_{max} conditions.

7. Chromatography as a Tool for Measurement of Analytes in Complex Mixtures

As was shown in Chapter 1 in Tables 1.2, 1.3, and 1.4, biological matrices are often complex mixtures. Given this complexity of biological matrices, interferences with the analyses of individual components in biological matrices are common. Thus it is a logical step to reduce the complexity of the matrix by separating the analytes from the interfering compounds to reduce or eliminate these interferences. The usual goal is to use a technique or a combination of techniques that result in all the analyte being in one fraction and all the other components of the matrix in a different fraction. In the separation processes, the differences in the chemistries of the analyte and the other components of the matrices are exploited. Some of the common separation processes include chromatography, extraction, precipitation, dialysis, ultrafiltration, distillation, electrophoresis, crystallization, cold trapping, electrodeposition, gasification, and centrifugation. Chromatography is one of the most frequently used separation techniques used by those who do chemical analyses in biological systems.

121

Chromatography encompasses a class of separation techniques based upon the affinity of an analyte for one physical phase over another physical phase. The technique was first described in 1906, and Tswett, its discoverer, named it chromatography because the separation of the plant pigments he was studying resulted in colored bands.

All chromatographic systems have two phases, a mobile phase and a stationary phase. The sample is loaded onto the stationary phase. Then successive volumes of mobile phase are passed over the stationary phase and the analytes are separated as a result of the differences of their "affinity" for the mobile phase and their "affinity" for the stationary phase. Analytes that have more "affinity" for the mobile phase move through the chromatographic system more rapidly, whereas analytes with more "affinity" for the stationary phase move more slowly. Figure 7.1 presents a schematic of the chromatographic process. The sample solution is loaded onto the top of the column and then a stream of the mobile phase starts flowing through the column. The compound with the most affinity for the mobile phase (as opposed to the stationary phase) will move the quickest through the column and be contained in the earlier portion of the column effluent; the compound with an intermediate affinity for the mobile phase will then be eluted in a later portion of the column effluent; and the compound with the highest affinity for the stationary phase will be eluted last.

Molecules of biological interest are separated or resolved by a variety of chromatographic systems. The most common chromatographic systems are gas chromatography (GC), thin-layer chromatography (TLC), and liquid chromatography (LC) (and its big brother, high-performance liquid chromatography,

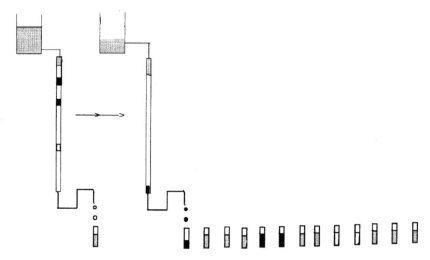

Fig. 7.1. A schematic of the chromatographic process.

HPLC); see Table 7.1. The predominant chemistries for the separations in each of these chromatographic systems are given in Table 7.2. The outputs of HPLC and GC systems are called chromatograms and generally have the same format. The detector response is plotted on the y axis, and time (or volume) is plotted on the x axis. A typical chromatogram is shown in Figure 7.2.

Chromatography can be either analytical or preparative in nature. Analytically, chromatography is used to analyze the composition of a mixture by separating the components of mixture from each other, detecting each component and then quantifying the concentration of each component. Preparative chromatography is used for the isolation and purification of compounds and will not be discussed further in this book.

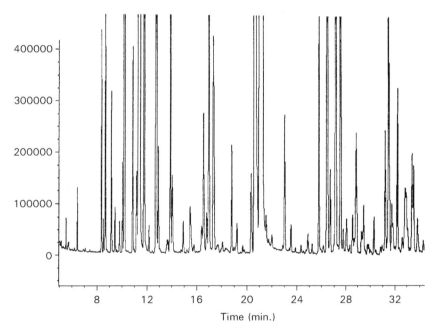

Fig. 7.2. A GLC chromatogram of an oregano extract. Each peak and each shoulder on each peak is most likely a unique compound. Frequently one peak can contain more than one compound.

THIN-LAYER CHROMATOGRAPHY

Thin-layer chromatography systems are operationally the simplest chromato-graphic systems. The mechanisms for separation are partition, ion exchange and adsorption. A TLC stationary phase consists of adsorbent-coated glass plates,

TABLE 7.1.   Types of Chromatography

Type	Solid Phase	Mobile Phase
Liquid chromatography (LC and HPLC)	Particles in a tube or coated tubing	Liquid
Thin-layer chromatography (TLC)	Coated plates or paper	Liquid
Gas–liquid chromatography (GC)	Particles in a tube or coated tubing	Gas

TABLE 7.2.   Predominant Chemistries in Each Type of Chromatography

Type of Chromatography	Separation due to
Liquid chromatography	
Ion exchange	Charge–charge interaction
Partition	Partition between liquid phases
Adsorption	Adsorption on solid surfaces
Affinity	Highly selective affinity for given biological molecules
Size exclusion (gel filtration)	Molecular sieving
Gas chromatography	Boiling point, vapor pressure, partition between liquid phases, adsorption on solid surfaces
Thin–layer chromatography	
Ion exchange	Charge–charge interaction
Partition	Partition between liquid phases
Adsorption	Adsorption on solid surfaces

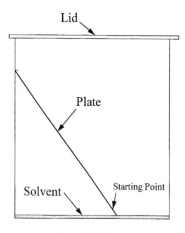

Fig. 7.3.   A TLC system.

TABLE 7.3. Absorbents used in TLC

Solid Support	Used to Separate
Silica gel	Amino acids, alkaloids, sugars, fatty acids, lipids, inorganic anions and cations, steroids, terpenoids
Alumina	Alkaloids, food dyes, phenols, steroids, vitamins, carotenes, amino acids
Kieselguhr	Sugars, oligosaccharides, dibasic acids, fatty acids, triglycerides, amino acids, steroids
Celite	Steroids, inorganic cations
Cellulose	Amino acids, food dyes, alkaloids, nucleotides
Ion exchange	Nucleotides, halide ions, amino acids
Polyamide powders	Anthocyanins, aromatic acids, flavonoids, proteins

adsorbent-coated plastic sheets or paper. Typical absorbents for thin-layer chromatography classified by the components to be separated are given in Table 7.3. The mobile phases in TLC can consist of almost any liquid aqueous and/or nonaqueous solvent system. There are literally thousands of different TLC systems, and a search of the literature will usually lead to a system for the separation of interest. A drawing of a TLC system is shown in Figure 7.3 and a typical chromatogram is shown in Figures 7.4a and 7.4b.

Operationally the samples of interest are spotted at one end of a chromatographic plate. The TLC plate is placed into a sealed development chamber containing the mobile phase.[1] The chromatogram is allowed to develop until the mobile phase front has moved some specified distance up the plate. The TLC plate is removed from the chamber, the mobile phase front is marked (often with a pencil), and the plate is dried. The spots are then visualized using the appropriate reagent and marked. The distance traveled by each spot is then measured. The distances moved in TLC systems vary a great deal due to differences in distance traveled by the mobile phase front, room temperature, and degree of saturation of the air space with the volatile components of the solvent system in the development chamber. However, the relative distances are reasonably constant. Thus in TLC relative distance traveled is reported. The relative distance traveled is usually referenced to the solvent front, and is called the R_f:

$$R_f = \frac{\text{distance traveled by the component of interest}}{\text{distance traveled by the solvent front}} \qquad (7.1)$$

For example, the R_fs of the compounds in the completed TLC chromatogram in Figure 7.4b are given in Table 7.4. It should be noted that R_fs of 0 and 1 are not

---

[1] *Note:* The initial level of the mobile phase should be below the samples spots on the TLC plate to prevent leaching of the samples into the mobile phase in the TLC tank.

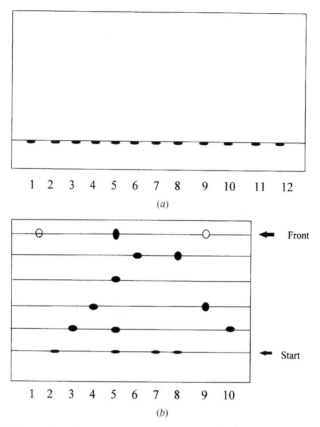

Fig. 7.4. (a) The spotting of a TLC prior to development. Lane 1—compound A; lane 2—compound B; lane 3—compound C; lane 4—compound D; lane 5—unknown 1; lane 6—compound E; lane 7—unknown 2; lane 8—unknown 3; lane 9—unknown 4, and lane 10—unknown 5. (b) The TLC plate after development and treatment with a reagent for the detection of the analytes. The data treatment for this chromatogram is given in Table 7.4.

considered to be very useful for the identification of compounds, since in the first case none of the compound moves with the mobile phase and in the second case all of the compound moves with the mobile phase. In either case there are frequently several compounds in a sample that will have such mobilities. When the R_f of an unknown compound in a sample is the same as that of some known compound and that R_f is greater than 0 but less than 1, then the similarity in R_fs is considered reasonable evidence (not proof) that the unknown analyte can be identified as the known compound. Occasionally different compounds in a sample will have different colors and their identification can be significantly aided

TABLE 7.4. **Results of Chromatogram Shown in Figure 7.3b**

Spot No.	Distance Moved (cm)	R_f	Identity	Color	Comment
1	8.0	1.0	Known A	Yellow	
2	0	0	Known B	Blue	
3	1.5	0.19	Known C	Blue	
4	3.1	0.39	Known D	Blue	
5	0	0	B?	Blue	Poor ID, no movement
	1.5	0.19	C	Blue	
	4.8	0.60	?	Blue	
	8.0	1.0	?	Blue	Not A, wrong color
6	6.5	0.81	Known E	Red	
7	0	0	B?	Blue	Poor ID, no movement
8	0	0	B?	Blue	Poor ID, no movement
	6.5	0.81	E	Red	
9	3.1	0.39	D		
	8.0	1.0	A?	Yellow	Poor ID, total movement, but color same as A
10	1.5	0.19	C	Blue	

by a comparison of the colors of the spots of the known compounds and the colors of the analytes in the unknown samples. Many compounds fluoresce under ultraviolet lamps, and their fluorescence can be used for their detection and identification.

In the example shown in Figure 7.4b, five known samples and five unknown samples were run. The interpretation of this run would be as follows. Spot 5 (unknown 1) has at least 4 components in it. One might be compound B, although that identification is not very trustworthy since the R_f is 0; the second would appear to be compound C; the third component has an R_f of 0.60 and cannot be identified because none of the known components has that R_f. The fourth is unknown and definitely not compound A, since the color is different from compound A. Other standards would have to be chromatographed to aid in the identity of the third component and the fourth component. Spot 7 (unknown 2) would appear to be compound B, although that identification is not very trustworthy, since the R_f is 0. Spot 8 (unknown 3) would appear to have at least two components; one would appear to be compound E because both the R_f and the color are the same as compound E; the second might be compound B, although that identification is not very trustworthy, since the R_f is 0. Spot 9 (unknown 4) has at least two components. One would appear to be compound D, since the R_f and the color are the same. The second might be compound A, since the color is the same; however, this identification is tentative, since the R_f is 1.00 and that is an untrustworthy R_f for identification.

TLC systems are usually rapid and require simple and inexpensive apparatus. A wide variety of detection reagents can be used (particularly with inorganic stationary phases). The necessary skills can be easily learned by people with relatively little technical training. A wide range of sample sizes are possible.[2] Many good TLC systems have high sensitivities, and typically micro- to nanogram amounts of analytes can be detected. TLC experimental parameters can be readily altered to effect separations. TLC is the only chromatographic system wherein samples with high affinity for the stationary phase are readily detectable. Multiple samples can be run on one chromatographic plate, and TLC is often used to monitor chemical reactions. When an analyst has many samples with relatively few components, TLC is usually the cheapest and quickest way to analyze those samples. TLC cannot be used for the analysis of volatile analytes, since they evaporate during the chromatogram development or in the detection phase. Most TLC systems are more qualitative than quantitative. TLC quantifying systems are expensive and usually have low precision. The developed original TLC plates are difficult to store, and most analysts photograph the plates and then discard them.

LIQUID CHROMATOGRAPHY

In liquid chromatography a tubular column is filled with a solid matrix with an accompanying mobile liquid phase. The sample is applied to the column, and the separation in chromatography occurs when successive volumes of mobile phases flow through the bed of the stationary phase. In liquid chromatography the mobile phases are various liquids and solutions and the stationary phases are either a variety of finely divided solids (preferably uniform-sized spheres) packed in a column or insoluble coatings on the column. For liquid chromatography to be considered for the separation of any mixture of analytes, the analytes *must* be soluble in the eluting mobile phases. The abbreviation LC is used to refer to the low-pressure liquid chromatography systems, and HPLC is used to refer to the high-performance liquid chromatography systems. The mechanisms of separation are the same for each and include molecular exclusion, ion exchange, partition, adsorption, and bioaffinity. It is not uncommon that more than one mechanism plays a role in the separation of individual analytes. Frequently the analytes are derivatized to change their solubility or their affinities for the stationary phase and/or mobile phase. While some liquid chromatography systems utilize a uniform mobile phase (isocratic elution), many more use changing compositions of the mobile phase. These concentration changes of the mobile phase can be done in discrete steps (step gradients) or in a wide variety of continuously changing concentrations (gradient elution). A typical HPLC system is shown in Figure 7.5, and an HPLC chromatogram is shown in Figure 7.6.

[2] Since multiple applications of samples can be loaded onto a TLC plate with drying of the spot between sample loads.

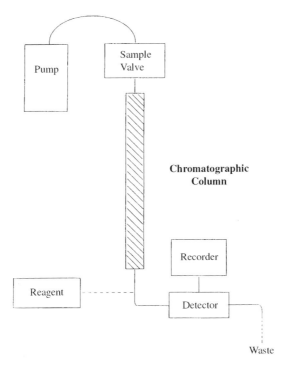

Fig. 7.5. An HPLC system.

In liquid chromatography the analytes are identified by their elution volumes and/or elution times,[3] and the amount of the analyte is determined from the intensity of the signal of each individual analyte. Given that real-world biological matrices can be very complex, many chromatograms have peaks that contain more than one component. Thus the identification of analytes by elution volumes always carries some uncertainty. Selective detectors and/or postcolumn colorimetric reactions can be used to improve the accuracy of the identification of the individual analytes. As shown in Table 7.5, there are numerous types of liquid chromatographic detectors. Refractive index and uv/vis spectrophotometric detectors are most commonly used.

SIZE EXCLUSION (GEL FILTRATION) CHROMATOGRAPHY

Gel filtration chromatography (size-exclusion chromatography, molecular sieve chromatography) separates solutes based upon the differences in their molecular

---

[3] Identification by elution time requires constant flow rates.

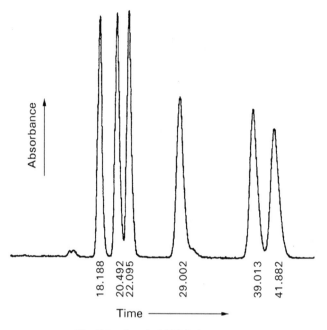

Fig. 7.6. A typical HPLC chromatogram.

size. The stationary support consists of porous gel materials that are filled with the mobile phase. The pores in such gels have diameters in the range of the Stokes radii of the solutes to be separated. Large solutes cannot penetrate any of the pores and thus travel quickly through the columns because they only have access to the excluded solvent. Small solutes can penetrate all the gels' pores and thus travel slowly through the column matrices because they have access to all the solvent. Medium-sized solutes can penetrate some of the gels' pores and thus

TABLE 7.5.   Common Liquid Chromatographic Detectors

Detector	Approximate Limit of Detection (g/mL)	Selectivity	Useable with Gradients
Refractive index	10^{-7}	Universal	No
Light scattering	10^{-4}	Universal	Yes
Conductivity	10^{-8}	Ions only	No
Absorbance (uv/vis)	10^{-10}	Selective	Yes
Electrochemical	10^{-12}	Selective	Difficult
Fluorometric	10^{-11}	Selective	Yes
Mass spectrometry	10^{-9}	Universal	Yes

travel at intermediate rates according to their Stokes radii, the larger molecules moving more rapidly and the smaller molecules moving slower. A typical gel filtration chromatogram is shown in Figure 7.7. The crucial parameters of gel filtration chromatography are defined in Table 7.6. For those solutes whose K_ds are between 0.9 and 0.1, gel filtration can be used to determine the molecular weight of the analyte. The gel filtration column is calibrated with the standards with analogous chemistries to the analyte (e.g., protein standards of known molecular weights of given shapes would be used to calibrate a column for the determination of the molecular weight of a protein). A plot of the elution volumes of the solutes of known molecular weight versus the logarithm of their molecular weights should give a linear plot. The molecular weight of the unknown analyte is then computed using its elution volume (see Figure 7.8). Gel filtration chromatography stationary phases are typically composed of beads of agarose or cross-linked dextrans (Sephadex), polyacrylamide, or porous glass. For best results it is very important that the beads be of uniform size, that the pore size distribution be Gaussian and cover the size range of interest. Gel filtration chromatography done with agarose, cross-linked dextrans, and aqueous buffers are known to be gentle chromatographic systems and are thus favored by those chromatographing fragile analytes such as easily denatured proteins. Gel filtration columns of agarose and cross-linked dextrans are susceptible to destruction at relatively low pressures, since the gels are not rigid and are rather easily crushed. On the other hand, the more rigid gels such as polyacrylamide and porous glass

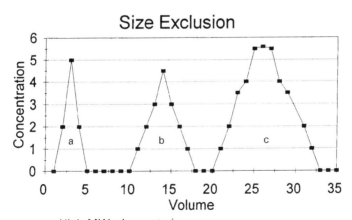

a. High MW - Largest size
b. Medium MW - medium size
c. Low MW - small size

Fig. 7.7. A gel filtration chromatogram.

TABLE 7.6. Special Terms in Gel Filtration Chromatography[a]

V_e	Elution volume of individual solute
V_0	Void volume (elution volume of a solute that can travel only in the liquid excluded from chromatographic gel)
V_t	Elution volume of a solute that can penetrate all parts of the available liquid (note this definition does not include the volume of the gel itself)
V_i	Total available liquid volume of column matrix $V_i = (V_t - V_0)$
K_d	Partition coefficient of analyte (always measured as a ratio of the elution of two solutes), $K_d = (V_e - V_0)/(V_t - V_0)$

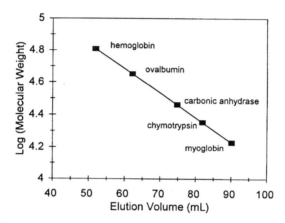

Fig. 7.8. Determination of molecular weight using gel chromatography.

beads, which can yield much faster chromatography, are much more susceptible to nonspecific binding, particularly with protein analytes.

ION-EXCHANGE CHROMATOGRAPHY

The interaction between ionic species in solution and ion-exchange resins is a competitive process and is correctly called ion exchange because one ionic species exchanges with another ionic species. The primary separation mechanism for ion-exchange chromatography is charge–charge interaction. (see Figs. 7.9–7.11. The ion-exchange chromatographic supports consist of various insoluble materials [typically polystyrene, cross-linked dextrans (Sephadex), cellulose, and porous glass (silica) beads] which have charged groups attached to them (see Table 7.7). There are two types of ion-exchange chromatography resins; cation-exchange and anion-exchange resins. A cation-exchange resin (e.g., carboxymethyl CM) has immobilized negative charges that can interact with molecules carrying a net positive charge, cations. An anion-exchange resin (e.g., diethylaminoethyl, DEAE), has an immobilized positive charge that can interact with molecules carrying a net negative charges, anions.

In aqueous solutions charged molecules do not exist by themselves. A positive charged molecule is always accompanied by a negatively charged molecule. For

TABLE 7.7. Functional Groups Used for Ion Exchangers

Anion Exchangers	Functional Group	Counter Ion
Aminoethyl- (AE-)	$-OCH_2CH_2N^+H_3$	Cl^-
Diethylaminoethyl- (DEAE-)	$-OCH_2CH_2N^+H[CH_2CH_3]_2$	Cl^-
Quaternary aminoethyl- (QAE-)	$-OCH_2CH_2N^+[CH_2CH_3]_2$	Cl^-
Cation Exchangers		
Carboxymethyl (CM)	$-OCH_2CO_2^-$	Na^+
Phospho	$-OPO_3^-H$	Na^+
Sulfopropyl (SP)	$-CH_2CH_2CH_2SO_3^-$	Na^+

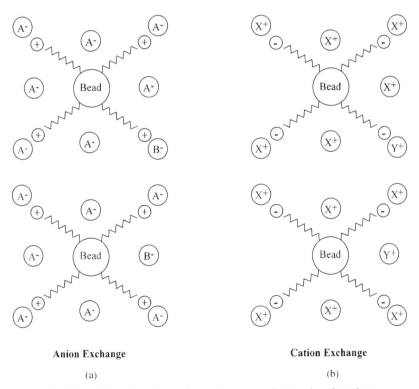

Anion Exchange

(a)

Cation Exchange

(b)

Fig. 7.9. Schematics of ion exchange chromatography (a) anion; (b) cation.

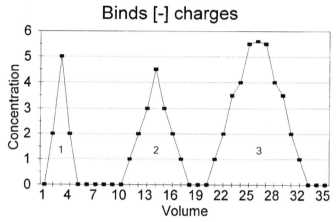

Fig. 7.10.   An anion-exchange chromatogram.

example, while the carboxymethyl group attached to the CM resin can have a "fixed" negative charge, it is always associated with a positively charged "counter ion" such as Na^+ or K^+. Thus the negative charge of a cation-exchange resin will interact with a buffer cation, an analyte cation, or a salt cation. The

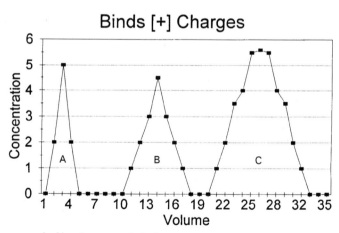

Fig. 7.11.   A cation-exchange chromatogram.

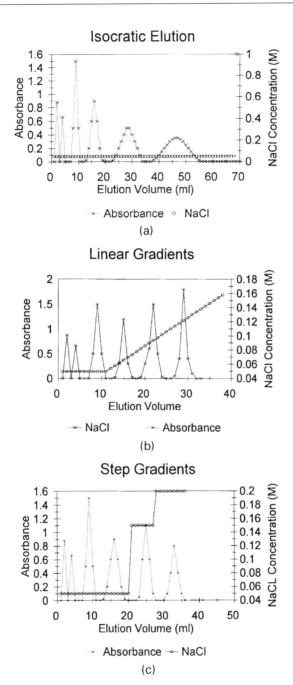

Fig. 7.12. Concentration gradients in ion exchange chromatography (a) Isocratic elution (b) linear gradients (c) step gradients.

cation that actually interacts with the negatively charged carboxymethyl group will depend upon several factors, including the net charge of the cationic species and its concentration.

The pH of an ion-exchange chromatographic system can have a significant effect on its separation capabilities. The charge on the ion-exchange matrix and on the analyte may well be dependent upon the pH of the chromatography system. For an ion-exchange resin to work, its functional groups must carry a charge (i.e., exist as a salt). In other words, for a CM resin the pH of the solution in which the resin is suspended must be high enough so that the carboxyl group is unprotonated (negatively charged) while for a DEAE resin the pH must be low enough so that the amino group is protonated (positively charged). Ionic strength will also affect the binding of the analyte to an ion-exchange resin. The analyte ion-exchange interactions are weakened by increasing the ionic strength of the eluting mobile phase (typically by increasing the concentration of NaCl or KCl).

In most ion-exchange chromatography assays, the pH, ionic strength, and temperature[4] of the system are chosen so that the analyte(s) of interest initially bind tightly to the column. The analytes are then eluted by systematically changing elution buffer, usually by increasing its ionic strength. The changes in ionic strength are done either in steps (step gradients) or in some continuous fashion, typically with a linear or exponential increase in ionic strength (linear or exponential gradients; see Fig. 7.12).

Some analytes such as proteins are very tightly bound to ion exchangers with polystyrene, polyacrylamide, and porous glass (silica) backbones, and losses can occur when such chromatographic materials are used. In such cases backbones of agarose, cellulose, and cross-linked dextrans are recommended.

REVERSE-PHASE CHROMATOGRAPHY

Reverse-phase chromatography can be viewed as a form of hydrophobic interaction chromatography in which hydrophobic (antagonistic to water, non-polar[5]) molecules (or molecules with hydrophobic areas on their surface) interact with each other rather than interacting with water molecules. In reverse-phase chromatography the stationary phase consists of hydrophobic molecules, such as long-chain hydrocarbons, bound to a silica matrix. The polar molecules (water, charged molecules, sugars, etc.) will not bind to the hydrophobic compounds and

---

[4]Remember that the pK_a of many acids are temperature dependent.

[5]Polar molecules are those that are charged in solution or have a separation of charge in solution. Nonpolar molecules are those that have no separation of charge within the molecule. Ionic compounds dissolved in water are polar, as is water. The order of polarity is approximately water > alcohols > ethyl ether > hydrocarbons. The longer a hydrocarbon chain in a molecule, the lower the polarity. The polarity of a solvent can be altered by adding a second component with a different polarity.

thus will remain in the aqueous mobile phase and be eluted in the initial fractions. Nonpolar or hydrophobic molecules in an aqueous mobile phase will selectively interact with the hydrophobic molecules.

The organic phases that are covalently bound to the silica gel include alkyl (typically $-Si-C_{18}H_{37}$ or $-Si-C_8H_{17}$) or alkyl nitriles ($-Si-C_2H_2-C_2H_2-CN$) or phenyl ($-C_6H_5$). The elution of the components from such reverse-phase chromatography systems is in order of decreasing polarity. Initially the mobile phases have a high polarity and then their polarity is decreased to elute the nonpolar compounds. The mobile phases typically consist of mixtures of water and acetonitrile of decreasing polarity.

ADSORPTION CHROMATOGRAPHY

Adsorption chromatography is based upon the partitioning of the analyte between the liquid components of the stationary phase and its adsorption to the stationary-phase material of the stationary phase. This was the type of chromatography that was developed by Tswett and is the earliest of all chromatography types. Typical stationary-phase materials include alumina, silica, and charcoal. There are many others. The mobile phases are typically mixtures of polar and nonpolar solvents. The usual procedure is to select a mixture that has a polarity approximately equivalent to that of the most polar analyte in the mixture.

AFFINITY CHROMATOGRAPHY

Many biological molecules have highly selective affinities for certain ligands. For example, enzymes have selective affinities for their substrates and competitive inhibitors, antibodies have selective affinities for their antigens, DNA sequences for their complementary sequences, avidin for biotin, some lectins for cell membranes, hormones for receptor proteins, etc. Affinity chromatography utilizes this selective affinity for the specific isolation of certain ligands. The affinity chromatographic support is synthesized by chemically binding (usually covalently) the selective ligand to a stationary matrix. The stationary matrix with its ligand is then used as a chromatographic support in liquid chromatography. When a mixture of biological molecules, such as a cell extract is applied to the column, the biological molecules with a selective affinity for the ligand will be either bound or retarded in their movement through the column. The desired biological molecules are then eluted with a solution of the ligand or some other compound which binds at the site where the ligand is bound. Alternatively, the eluting compound has the ability to alter the binding of the ligand to the biological molecule of interest. For example, changes in pH or ionic strength could alter the binding. Common bound ligands include competitive enzyme inhibitors (e.g. substitute analogs), substrate analogs, antigens, antibodies, commentary RNA or DNA sequences, enzyme cofactors, as well as other compounds which have highly selective binding sites on the biological molecules of interest. Eluting agents include the same list of compounds plus alteration of pH and ionic strength. Affinity chromatography is normally used in isolation procedures and is rarely used in an analytical mode. Enrichments of several hundred fold are not uncommon. Occasionally the affinity chromatography can be used to do a one step purification of the component of interest.

GAS CHROMATOGRAPHY

A drawing of typical gas chromatographic[6] systems is shown in Figure 7.13. In gas chromatography the stationary phase consists of solid particles packed inside a tubular column or a film coated on the inside of the tubular column. In GC the mobile phase is volatile and the analyte must also be volatile. Frequently the analytes are derivatized to improve volatility. The mechanisms of separation for GC are volatility, adsorption chromatography and partition chromatography. Packed columns were originally popular and are still used for the separation of simple mixtures. However, the development of coated-capillary columns has led to GC systems with very high resolving power. Modern coated-capillary columns have the capability of resolving 50 to 100 different compounds in one chromatographic run (see Fig. 7.2) at limits of detection of approximately one nanogram per injected sample. Most GC detectors are based upon either the ionization of the analyte or the carrier gas or upon changes in the thermal conductivity of the gas mixture passing through the detector. The hydrogen flame detector (ionizes the analyte) is by far the most common. It has very low limits of detection (approximately 1 ng), is relatively safe, and is much more popular than the ionization of argon, which requires a radioactive source. Electron-capture detectors are popular, since they only detect halogen compounds, nitriles, and nitrates. Since the electron-capture detectors do not respond to compounds that do not have these functional groups, they can be used to detect and quantify halogen compounds, nitriles, and nitrates even when such compounds are not separated from the other components of the matrix during chromatography.[7] Some specialized flame photometric detectors can be used to detect those compounds that contain phosphorus and sulfur. There is an increasing use of mass spectrographs as detectors in gas chromatography. Typically the mass spectrography is set to monitor only one or a very few masses, and thus the use of

[6] Sometime gas chromatography is called gas–liquid chromatography (GLC).

[7] For example, in the analysis of chlorinated pesticides in environmental samples.

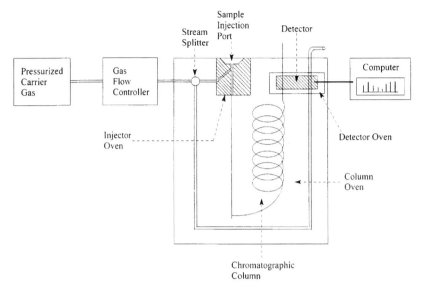

Fig. 7.13. A schematic of a typical GC system.

the mass spectrograph can add significant selectivity to a gas chromatographic run.

INTERNAL STANDARDS

It is difficult to reproducibly load small sample volumes into HPLC or GC columns. Variations in the solvent compositions and/or temperature of the chromatographic column and differences in injection volumes can cause differences in elution times and in the relative responses of the analytes. Many post-column detection reagents decay over time. All these variables can cause deterioration in the precision and/or accuracy of HPLC and GC assays. Internal standards are known compounds that are added to the analyte sample in known concentrations. Chromatographers use internal standards to validate elution times and reaction chemistries and to correct for variations in injection volumes. It is critical that the internal standards do not elute at the same time as any other components of the sample, and it is desirable that internal standards have the same general chemical characteristics as the analytes with respect to both chromatographic behavior and detection behavior. Most chromatographers believe that the failure to use internal standards is a serious error, and often they will distrust any chromatographic data obtained without the use of internal standards.

ANALYZING A GC OR HPLC CHROMATOGRAPHIC RUN

While there are obvious differences between GC and HPLC, the chromatograms from the two systems are analyzed in the same manner. All chromatographic assays are empirical in nature. Thus each chromatographic system must be independently calibrated. In the initial calibration step known amounts of individual standards are injected to determine the elution times of each compound. The second step is to inject mixtures of known composition and concentration (including an internal standard) to determine the responses of the chromatography system to each compound. Finally, one chromatographs the test samples, and utilizes the data from the chromatography of the known compounds to identify and quantify the analytes in the unknown sample. There are two steps in the analysis of a chromatogram: identify and quantify. Thus the process might be called a chromatographic IQ: Identify, then quantify.

Identification

The individual compounds in a chromatogram are identified by their elution volumes.[8] These elution volumes are compared with those of known compounds and must be the same within experimental error. The validation step of this process is the cross-check that the internal standard has its previously determined elution volume. Given that real-world biological matrices can be very complex, many chromatograms have peaks that contain more than one component. Thus the identification of analytes by elution volumes always carries some uncertainty. Selective detectors and/or postcolumn colorimetric reactions can be used to improve the accuracy of the indentification of the individual analytes.

Quantification

The first step in determining the amount of a component in a given peak is to measure the peak height or the peak area of the analyte chromatographic peak. This can be done manually by using a ruler to determine the height of the component peak above the base line or by determining the peak area by multiplying the measured peak height times the measured peak width at half-height. While such manual measurements will work, they are tedious, and their use leads to results with poor precision. The preferred method for determining peak area is to use automated integration. With an automated integration system,

---

[8] Or times, if there is a constant flow rate.

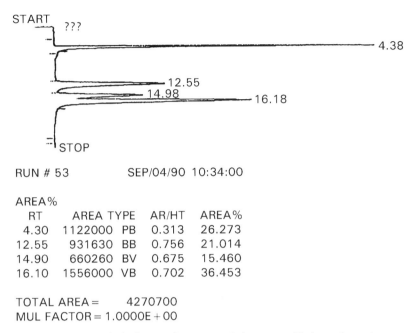

Fig. 7.14. A typical print out of an automated chromatographic integration system.

the area of the peak is determined automatically during the chromatographic run and typically printed out at the end of the run (see Fig. 7.14).

Area Ratio Technique

One of the best systems for computation of the initial concentration of the analyte in the sample is to use the area ratio technique. The basic concept in the use of the area ratio technique is that the ratios of the amounts of analyte to the internal standard stay the same when the aliquots were taken from the same sample no matter what volumes have been used as a chromatographic sample and what variations in the assay conditions have occurred. With the area ratio technique all samples (both standards and unknowns) are prepared such that the internal standard concentration is *identical* in both the standard mixture of known concentration and the unknown samples. The standard mixtures of known concentrations and unknown sample mixtures are chromatographed. Each component is identified by its elution time and then the peak area of each component is determined. Then the *area ratios* for each analyte in the standard mixture and unknown sample solution are computed. Area ratios are computed for each individual chromatogram, both the chromatograms for the standards of known composition and concentration and the chromatograms of unknowns of unknown composition and concentration:

$$\text{area ratio} = \frac{\text{area measurements for individual analyte}}{\text{area measurements for internal standard}} \qquad (7.2)$$

Then concentration of each identifiable analyte in the unknown samples is calculated:

$$\text{mg/mL of analyte in sample} = \frac{\text{area ratio for analyte in unknown sample}}{\text{area ratio for analyte in standard}}$$
$$\times \text{ mg/mL of analyte in standard} \qquad (7.3)$$

The area ratio technique is simple and straightforward and is used widely when the concentration of the internal standard in the injected sample can be held constant. Note that with the area ratio technique of computation the normalization steps of the analysis of chromatographic runs are an integral part of the computations.

Sample Calculations for HPLC or GC Chromatograms: Problem

In the analysis of two unknowns for taurine (an amino acid) 3 mg of tissue from each sample has been suspended in 2.0 mL of buffer containing 1.0 μmol/mL phosphoserine as an internal standard. In each case 10 μL of the solution was injected the HPLC column. The resulting chromatograms are shown in Figure 7.15. The known standard samples contained 1 μmol/mL of phosphoserine and 1 μmol/mL of taurine, and the results of the chromatographic analyses of the standard samples are shown in Table 7.8.

Unknown 1

1. **Identify.** The first peak elutes at 4.52 min and the second elutes at 12.16 min. There are no other peaks. The first peak must be phosphoserine, since it was present in the extraction buffer. The second should be taurine, since the difference in elution time of 7.64 min is the same as 7.49 min (e.g. 11.84 min minus 4.32 min, Table 7.8) within the measurement error of this system.

TABLE 7.8. **Data from Analyses of Known Standards**

Standards Amino Acid	Elution Time (min)	Area Ratio
Phosphoserine	4.32	—
Taurine	11.81	1.081

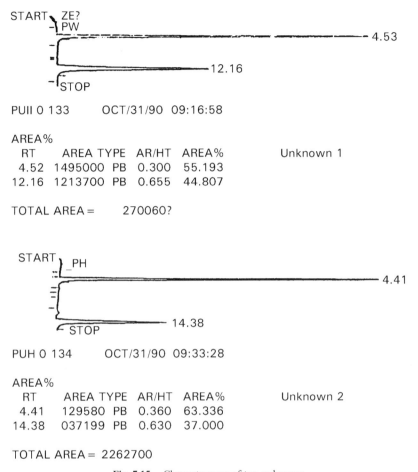

START ZE?
  -\PW
                                                              - 4.53
  -

  ■
                                    -12.16
  -⌐
   ┗STOP

PUII 0 133   OCT/31/90 09:16:58

AREA%
  RT   AREA TYPE AR/HT AREA%   Unknown 1
  4.52 1495000 PB 0.300 55.193
  12.16 1213700 PB 0.655 44.807

TOTAL AREA =   270060?

START) _PH
  ∷
  ≡   - 4.41
  -

  ⌐ STOP   - 14.38

PUH 0 134   OCT/31/90 09:33:28

AREA%
  RT   AREA TYPE AR/HT AREA%   Unknown 2
  4.41   129580 PB 0.360 63.336
  14.38 037199 PB 0.630 37.000

TOTAL AREA = 2262700

Fig. 7.15. Chromatograms of two unknowns.

2. **Quantify.** The area ratio of unknown 1 for taurine = 1219700/1495000 = 0.812. The concentration of taurine in unknown 1 is computed as follows:

$(0.812/1.081) \times 1\mu mol/mL = 0.75\mu mol/mL$;
thus there were 1.5 μmol per 2 mL or 3 mg of tissue. Thus there were 0.5 μmol of taurine/mg of tissue.

Unknown 2

1. **Identify.** The first peak elutes at 4.41 min, and the second elutes at 14.38 min. There are no other peaks. The first peak must be phosphoserine, since it was present in the extraction buffer. The second cannot be taurine, since the difference in elution time of 9.97 min is not the same as 7.49 min within the measurement error of this system.

2. **Quantify.** If there was taurine in this sample, then there should be a peak at about 11.90 min. There is not. Thus if there is any taurine in this sample, it is below the limit of detection, which is not given for this system. Since an injection of 10 μL of the internal standard is equivalent to 10 nmol of internal standard, a ballpark guess for the limit of detection is less than 0.1 nmol.

Color Value Computation Techniques

There are times when the area ratio technique cannot be used because the concentrations of the internal standard in the injected sample cannot be held constant (e.g., when an analyte must be extracted from complex matrices). In such cases the more complex color value computation technique can be used. Again known amounts of the analytes and internal standards are chromatographed and the respective peaks are measured.

Color values are computed for each analyte as well as the internal standard, where the color value is defined as:

$$\text{color value} = \text{peak area} / \text{amount of analyte injected} \qquad (7.4)$$

Then the test samples containing a *known amount of internal standard* are chromatographed, and the individual peaks identified and measured. The amount of the individual analytes is computed as shown. Since the analytes and internal standards are dissolved in a common solution, the *ratios* of amounts of the internal standard to the amount of analyte stay the same no matter what manipulations have occurred with the sample. Thus the *amount* of the analyte in the original sample can be computed as:

$$\frac{\text{amount of analyte injected}}{\text{amount of internal standard injected}} = \frac{\text{amount of analyte in sample}}{\text{amount of internal standard in sample}}$$

$$(7.5)$$

Therefore:

amount of analyte in sample

$$= \frac{(\text{amount of analyte injected})(\text{amount of internal standard in sample})}{\text{amount of internal standard injected}}$$

This use of internal standards can be very useful for assays of heterogenous mixtures of analytes in complex matrices where exact volumes and weights are difficult to determine.

SELECTION OF CHROMATOGRAPHIC SYSTEMS

Selection of the components of any chromatographic separation system depends upon several factors, including the chemistry of the stationary phase, the chemistry of the mobile phase, the chemistry of the analyte mixture, the cost of the analyses, the complexity of the matrix, and the number of samples. An overview of the applicability of GLC, HPLC, and TLC for different analyte–sample mixtures is given in Table 7.9.

A few simple rules will aid the analyst in choosing the appropriate chromatography system. The first item to check is the volatility of the analyte. The analyte must be volatile to be separated by GLC and cannot be volatile with thin-layer chromatography, since it would evaporate from the TLC plate. With HPLC the analyte must be soluble in the mobile phase or it will not move through the chromatography column. Charged and high-molecular-weight compounds have very low volatilities, and thus GLC is not an acceptable means of separating

TABLE 7.9. Selection of Different Classes of Chromatography

	GLC	HPLC	TLC
Volatile	Yes	Maybe[a]	No
Nonvolatile	No	OK[a]	OK
Charged	No[b]	OK	OK
High MW	No[b]	OK	OK
Many samples	Maybe[c]	Maybe[c]	Yes
Good quantitation	OK	OK	No
Resolution	Excellent	Good	Poor
Low cost	No	No	Yes
Complex mixture	Yes	Yes	No

[a] The question is not whether or not the analyte is volatile; it is whether the analyte is soluble.

[b] Neither charged compounds nor high-molecular-weight compounds are very volatile.

[c] Since these samples will have to be run one at a time, this could be expensive.

such compounds. HPLC and GLC have very high potential resolving powers and are often used to separate complex mixtures of multiple components. TLC has a fairly low ability to resolve similar compounds, and lacks the ability to separate mixtures with a large number of compounds (> 20); thus it is not a good technique for the separation of complex mixtures. Analysts frequently resort to various chemical tricks to overcome the fundamental shortcomings of the resolution capabilities of various chromatography systems. For example, while fatty acids have low volatilities, their esters have a much higher volatility, and thus the analysis of fatty acid mixtures frequently utilizes an esterification step so that the high resolving power of GLC can be used in their analysis.

8. Electrophoresis and Other Electrokinetic Separations

INTRODUCTION

Proteins, RNA, DNA, and polysaccharides are the primary classes of macromolecules in biological systems, and within each class there are an enormous number of individual macromolecules. A conservative estimate for the number of different proteins would be somewhere between ten and a hundred thousand different individual proteins in a mammalian species. It is very difficult to distinguish between most of these molecules using traditional techniques of solubility or the more modern techniques such as HPLC. As has been mentioned before, some excellent specialized techniques, including affinity chromatography and various immune assays, can show elegant selectivity for some individual macromolecules. However, neither affinity chromatography nor the various immune assays are useful as general techniques for the separation of a wide variety of macromolecules at the same time. Today the techniques of choice are those that are based upon separation techniques that exploit the size, shape, and charge of the macromolecules of interest. Currently, the methods of choice are all based upon electrokinetic separations.[1]

---

[1] Electrokinetic mobility is defined as movement of ions in solution due to potential applied across the media in which the ions exist.

THEORY

When electrical potential is applied across a solution as is shown in Figure 8.1, the ions in solution carry the current. Cations (positively charged ions) move toward the cathode, and anions (negatively charged ions) move toward the anode. If E is the potential of the electrical cell, q is the charge on the ion, and f is the frictional force on the ion, then the velocity of each ion is given by:

$$V = Eq/f \tag{8.1}$$

The mobility of the ion, μ, is given as:

$$\mu = V/E = q/f \tag{8.2}$$

The frictional force f is proportional to the radius of the ion, and the radius of the ion is proportional to the cube root of the molecular weight for *molecules of similar shape*. Thus the mobility is a function of the molecular weight of the ion:

$$\mu = kq/(\mathrm{MW})^{1/3} \tag{8.3}$$

where k is a proportionality constant dependent upon the molecular shape. Larger net charges and smaller sizes usually result in a greater mobility of the ions. These equations hold for all electrical cells and have led to the use of electrokinetic separations in a variety of systems. We will discuss the

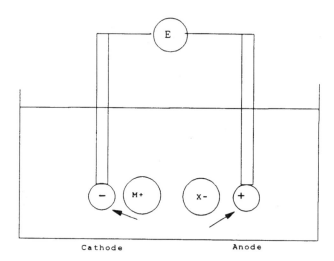

Fig. 8.1.   Movement of ions in a liquid electrical cell.

electrokinetic separations that are particularly useful for separating proteins and nucleic acids.

Free-Solution Electrokinetic Separations

Electrokinetic separations do occur in free solution; however, when the current and voltage are turned off, thermal gradients and molecular diffusion usually destroy much of the separation that might have taken place. Since the visualization process for electrophoretograms usually requires that the current and voltage be turned off, free-solution electrokinetic separations are rarely used in modern laboratories making chemical measurements in biological systems.

Electrokinetic Separations in Gels

An ideal support matrix will limit convection and diffusion, will not react with the molecules to be separated, will be easy to handle, and will have controllable porosity. A variety of support or stabilizing matrices have been used for electrokinetic separations, including paper, cellulose acetate, cellulose nitrate, starch, silica gel, polyacrylamide, and agarose. All these matrices limit convection and diffusion. However, only polyacrylamide and agarose also have controllable three-dimensional porous gels, which permit their use as molecular sieves. Most electrokinetic separations are done in supporting gels of polyacrylamide or agarose.

Definition: Electrophoresis is the separation of ions due to their differential mobility in an electric field.

Gel Electrophoresis

Gel electrophoresis is the most commonly used electrokinetic separation, and the most commonly used three-dimensional porous gels are polyacrylamide[2] and agarose. Polyacrylamide gels are usually used for electrokinetic separations of proteins and also are used for the separation of smaller nucleic acid polymers (DNA fragments[3]). Agarose gels are usually used for the electrokinetic separations of RNA and DNA. When polyacrylamide or agarose gels are used as electrophoretic support media, the techniques are called polyacrylamide gel electrophoresis (PAGE) or agarose gel electrophoresis, respectively.

---

[2] Starch was originally the polymer gel of choice for protein electrophoresis, but studies showed that acrylamide gels gave better resolution and more reproducible results.

[3] The DNA fragment sizes should be in the hundreds of base pairs if PAGE is used for their separation.

Operation of a Gel Electrophoresis System

While there are differences between specific techniques, there are a number of steps common all electrophoresis: (1) make the gel; (2) apply the sample to the gel; (3) apply the electric field and run the electrophoresis; (4) detect the analytes after electrophoresis. We will first discuss the electrophoresis of proteins. The electrophoresis of DNA molecules will be discussed later in this chapter.

PROTEIN ELECTROPHORESIS

Making the gels

The gel matrix is chosen with regard to its lack of chemical interaction with the analyte and its porosity. The gel is prepared in a buffer that is chosen based upon the planned electrophoretic separation. In most cases, wells for multiple samples are produced by inserting a well-forming "comb" in the gel-forming solution before polymerization has occurred. The gel polymerizes around the teeth of the comb to form the wells. After the gel is made, the gel and its support are placed inside the electrophoresis apparatus (see Fig. 8.2).

Polyacrylamide Gel Electrophoresis

The polyacrylamide gel matrix is prepared by a radical-initiated polymerization. In the presence of free radicals, acrylamide will polymerize to form linear polymers, but when bis-acrylamide (N,N'-methylene bisacrylamide) is added, it serves as a cross-linking agent, and three-dimensional matrices are formed. The commonly used radical-initiating reactions involve the use of ammonium persulfate (APS) and tetramethylethylenediamine (TEMED) or riboflavin, TEMED, and light. Figure 8.3 shows the structures of the monomer acrylamides, and Figure 8.4 shows a schematic for the polymerized gel. The polyacrylamide gel matrix has pores or holes, and the size of these pores is dependent upon the

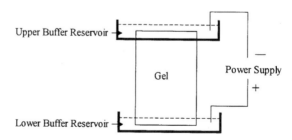

Fig. 8.2. An electrophoresis system. PAGE electrophoresis systems are usually vertical and agarose systems are usually horizontal.

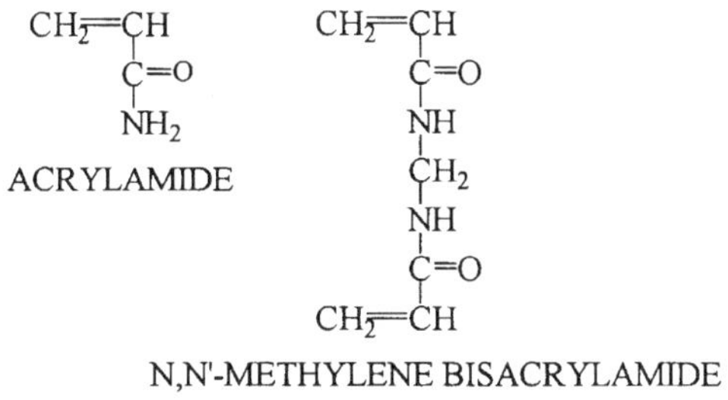

$$CH_2=CH \qquad CH_2=CH$$
$$\begin{matrix} | \\ C=O \end{matrix} \qquad \begin{matrix} | \\ C=O \end{matrix}$$
$$\begin{matrix} | \\ NH_2 \end{matrix} \qquad \begin{matrix} | \\ NH \end{matrix}$$
$$\text{ACRYLAMIDE} \qquad \begin{matrix} | \\ CH_2 \end{matrix}$$
$$\begin{matrix} | \\ NH \end{matrix}$$
$$\begin{matrix} | \\ C=O \end{matrix}$$
$$CH_2=CH$$

N,N'-METHYLENE BISACRYLAMIDE

Fig. 8.3. Structures of acrylamide and bisacrylamide.

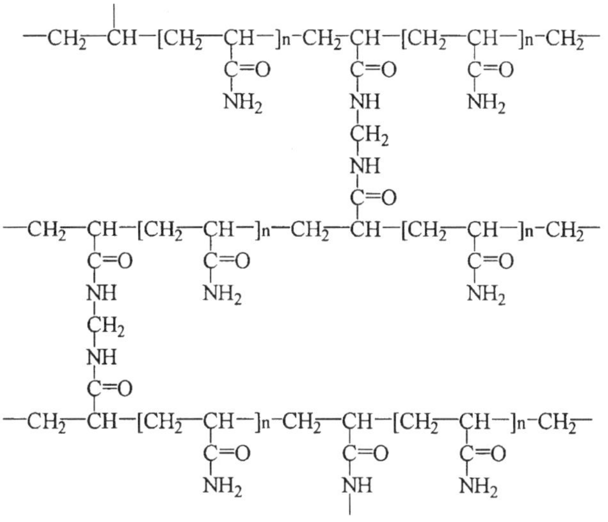

$$-CH_2-CH-[CH_2-CH-]n-CH_2-CH-[CH_2-CH-]n-CH_2-$$

Fig. 8.4. Polyacrylamide. This is a two dimensional representation of a three dimensional structure.

concentrations of acrylamide and N,N'-methylene bisacrylamide in the polymerization reaction. Higher concentrations of acrylamide result in smaller pore size. The acrylamide concentration of the polyacrylamide gel is usually reported in terms of the percent (w/v) of the acrylamide in the polymerization reaction (e.g., 8%). In terms of pore size, an 8% acrylamide gel will have larger pores than a 10% acrylamide gel. There are a number of common "recipes" for the preparation of polyacrylamide gels (see Table 8.1). In most cases, the ratio between the concentrations of acrylamide and N,N'-methylene bisacrylamide is fixed, and the pore size of the polyacrylamide gel is altered by using more or less of this acrylamide-N,N'-methylene bisacrylamide solution in the polymerization reaction. The pores in any polyacrylamide gel vary in size as a statistical function around an average pore size; some pores are larger than the average, and some are smaller. The polymerization mixture also includes a buffer system that can be varied as required (see below). The choice of a radical initiator depends upon the stability of the molecules to be separated. For example, some proteins are adversely affected by the byproducts of the ammonium persulfate–dependent system. In this case, the riboflavin-dependent system should be used. The rate of the polymerization process can be controlled by altering the concentration of ammonium persulfate (or riboflavin) or TEMED. Increasing the concentration of either reagent will speed polymerization, while decreasing the concentration will slow polymerization. Ideally, polymerization should be slow enough so that the gels can be poured before the mixture polymerizes, but fast enough so that the polymerization is complete within a reasonable time.

Polyacrylamide gels are commonly produced (poured) in one of two formats: slabs or tubes. The basic principles of electrophoresis apply equally well to both types of gel. With slab gels multiple samples can be applied to each gel, while only one sample can be applied to each tube gel. The tube gels originally were predominantly used, but recently analysts have tended to use the slab gels probably because the available technology makes slab gels easy to make and run and because comparison of the mobility of multiple samples is much more easily done with slab gels than on tube gels. There are several commercial kits for preparing slab gels of differing thicknesses, sizes, and well numbers. Precast gels are also commercially available.

Continuous and Discontinuous PAGE

The two common types of gel electrophoresis can be classified as continuous electrophoresis and discontinuous electrophoresis (disc electrophoresis). In each

TABLE 8.1. PAGE Separating Gels

Solution	Percent Gel				
	15%	12.5%	10%	9%	8%
A (mL)[a]	15	12.5	10	9	8
B (mL)[b]	7.5	7.5	7.5	7.5	7.5
dH$_2$O (mL)	7.3	9.8	12.2	13.2	14.15
APS (μL)[c]	200	200	300	300	350
TEMED (μL)[d]	13	13	13	13	13
Total vol. (mL)	30	30	30	30	30

[a]Solution acrylamide. Acrylamide, 30.0 g; Bis-acrylamide 0.8; dH$_2$O, to 100 mL; Degas-wrap in aluminum foil.

[b]Solution B: PAGE Running Gel Buffer, 1.5 M Tris-HCl, pH 8.9. Tris 18.17 g; dH$_2$O, to 100 mL; adjust to pH 8.9 with HCl.

[c]APS catalyst (10% ammonium persulfate). Ammonium persulfate, 0.1 g; dH$_2$O, to 1.0 mL.

[d]TEMED, N,N,N′N′-tetramethylethylenediamine.

the separation is made based upon the size and charge of the molecules under the conditions of the assay. They differ in the continuity of buffers in the separation systems. A continuous polyacrylamide gel electrophoresis system consists of a polyacrylamide gel with a single gel density (acrylamide concentration) and a single pH buffer (both the electrode buffer and gel buffer). The final band widths of the individual components are always equal to or greater than the band width of the applied sample, and the resolution between similar ions is poor. The sample application is critical and is frequently the limiting factor in the resolution of the system. The limitations of continuous electrophoresis are overcome when the band-sharpening disc electrophoresis systems are used.

Discontinuous Electrophoresis

In disc electrophoresis the width of the sample zone is relatively unimportant since the sample bands are sharpened in the process of the electrophoresis. The system works by using two buffers and two or three types of gel in a single electrophoresis system. A diagram of such a system is shown in Figure 8.5. In a typical system the sample buffer and the electrode buffer would be pH 8.3 $Tris^+glycine^-$, and the running gel[4] would be prepared with a running buffer of pH 7.3 $Tris^+Cl^-$. The sample gel and the stacking gel would be large-pore gels, and the running gel would be a small-pore gel. The disc electrophoresis systems are normally run at a fixed current.

Understanding the disc electrophoresis system requires an understanding of the concept that the amount of current carried by an ion species is dependent upon its mobility and its concentration. The mobility of an ion in an electric field is dependent upon its size, shape, and net electric charge. The smaller the ion, the greater its mobility. Given equal size, the ion with the greater net charge will have the greater mobility. The negatively charged current-carrying ions for a disc electrophoresis buffer system used in these experiments are chloride, glycine, and the proteins in the samples; the positive ions are protonated Tris (Fig. 8.6). The chloride ion has a much greater mobility than the glycine ion, and most the proteins have a mobility intermediate between these two ions. (The buffer pHs are chosen so that the proteins are negatively charged.)

The mobility of ions in the electrophoresis system is dependent upon their net charge and, at least in the separating gel, their size. The ions that migrate into the gel are negatively charged anions, since the positive pole (anode) is located in the lower buffer reservoir. The speed of migration is dependent upon the voltage applied to the system. Since protein electrophoresis is normally run under conditions of constant current, the actual voltage experienced by an ion is dependent upon the local resistance. To understand the stacking of the sample,

---

[4] Sometimes the running gel is called the separating gel.

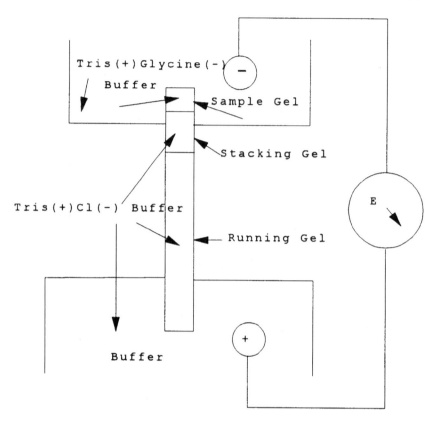

Fig. 8.5. A disc electrophoresis system. Note, the cathode buffer could also be the tris-glycine buffer.

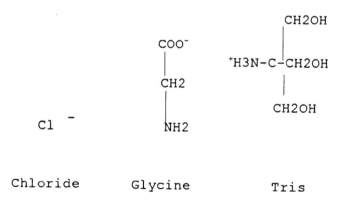

Fig. 8.6. Major current carrying ions in an disc electrophoresis system.

consider the Cl^- and/or phosphate[5] ions present in the stacking- and separating-gel buffers, the proteins, and the glycine present in the electrode buffer. The Cl^- and phosphate ions are small, and under the conditions that exist in the stacking and separating gel will always carry a negative charge. Proteins that separate in the gel system must carry a negative charge. The fraction of glycine that exists as the zwitterion with no net charge or as the glycinate anion with a net negative charge varies during the electrophoresis since the charge on the glycine is pH dependent.

In the electrode buffer, pH 8.3, glycine [$pK(\alpha\text{-COOH}) = 2.34$; $pK(\alpha\text{-NH}_3^+) = 9.60$] carries a net charge of -0.048. However, once the electric field is established and the anions migrate toward the anode, the glycine (and proteins) enter the stacking gel. The pH in the stacking gel, 7.2 or 6.8, is such that the net charge on glycine is decreased to -0.004 or -0.002. The Cl^- and phosphate ions continue to migrate in the electric field, but the migration of glycine is decreased because it has a diminished negative charge. As the Cl^- and phosphate ions move through the stacking gel, they leave a relative ion deficiency behind them. This deficiency increases the local electrical resistance. Since electrophoresis proceeds under conditions of constant current, this increase in resistance results in an increase in local voltage in the region between the leading Cl^- and phosphate ions and the trailing glycine. In the stacking gel, the protein charge-to-mass ratios and their mobilities are intermediate between that of the Cl^- and/or phosphate ions and glycine. The proteins experience a relatively large voltage in the stacking gel and move rapidly. The stacking gel has large pores that do not restrict the migration of proteins so that the proteins move as a group. The proteins are forced into a very narrow band, or stack, because of the local conditions they encounter. If the proteins move toward the leading Cl^- and phosphate ions, they experience decreased electric field and slow down. If the proteins lag behind, they experience an increased electric field and speed up. In the stacking gel, these opposing effects on the proteins, decreasing mobility near the Cl^- and phosphate ions and increasing mobility near the glycine, and the limited restriction of the large pore size result in the stacking phenomenon.

When the proteins encounter the smaller pores of the separating gel, their migration is slowed. Glycine overtakes the proteins at this point and enters the higher pH in the separating gel. In the separating gel, pH 8.9 or 8.8, the net charge on glycine increases to -0.17 or -0.14. This transition eliminates the ion deficiency experienced in the stacking gel and the electrophoresis proceeds under conditions of constant electric field strength. The Cl^- and phosphate ions precede the glycine molecules which in turn precede the protein anions.

Native Polyacrylamide Disc Gel Electrophoresis (Native-PAGE) of Proteins

Native-PAGE is the electrophoretic separation of proteins in their native shape and charge at the pH of the separation. Different proteins have different shapes

---

[5] Phosphoric acid is sometimes used to adjust the pH of electrophoresis buffers.

and charges and thus will have differing mobilities in the running gel. The most commonly used native-PAGE system uses separating and electrode buffer at high pH (~ 9). At this pH most proteins have a net negative charge and will move toward the anode.[6] Generally proteins with larger net charges move faster than proteins with smaller net charges. However, the rate of movement of the protein ions in the solution within a gel matrix with sieving properties is inversely proportional to the hydrodynamic shape of the protein molecules. These factors, net charge and size/shape, contribute to the movement of proteins with a polyacrylamide gel, and thus prediction of the relationship of movement of one individual protein species to that of another individual protein species is quite complex. While it is known that size and charge both contribute to the mobility, it is difficult to factor out the contributions of each. Thus native PAGE are rarely used for molecular-weight determinations. However, SDS-PAGE is often used for estimation of protein molecular weights (see below).

Native-PAGE is frequently used to evaluate the purity of an enzyme isolated from a complex mixture of proteins. This technique is also often used to identify isozymes (enzymes that catalyze a common reaction but that have different charge/size characteristics). In the case of the evaluation of enzyme purity a dual staining process is used (see below), one for the detection of all proteins and one for the enzyme activity. In this case, a single protein staining band that has equivalent mobility to a catalytically active band is consistent with a preparation that has one protein species that has the desired activity. Multiple protein stained

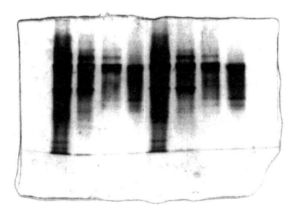

Fig. 8.7. A protein stained gel of a series of samples from an enzyme purification process. From left to right; a crude preparation of alchohol dehydrogenase, a partially purified preparation, a further purified fraction, and a pure fraction. The sequence is then repeated. See Figure 8.8 for the activity staining of the same preparations.

---

[6] There are other electrophoresis methods for the separation of proteins with large positive charge (e.g., histones).

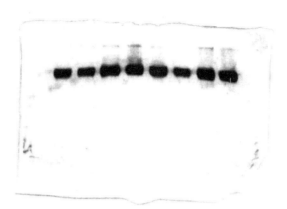

Fig. 8.8. A activity stained gel of a series of samples from an enzyme purification process. See Figure 8.7 for the protein staining of the same preparations. From left to right; a crude preparation of alchohol dehydrogenase, a partially purified preparation, a further purified fraction, and a pure fraction. The sequence is then repeated. Note that there is only one band with enzymatic activity and that it corresponds to the major protein band of the purified preparation. There also appears to be a minor protein band that does not have enzymatic activity in the most purified preparation.

bands with a single enzyme activity band would be an indication of an impure preparation (see Figs. 8.7 and 8.8).

There are a number of cases where there are multiple electrophoretic forms of single-enzyme activities. Such behavior is good evidence for the presence of isozymes. There are quite a few cases where isozymes are found in a single biological system. For example, humans have multiple electrophoretically-distinguishable forms of lactate dehydrogenase.[7] The ability to stain for enzymatic activity in these polyacrylamide gels requires that the enzyme maintain its native, active confirmation during electrophoresis and accounts for the procedure name, native-PAGE.

SDS (or Denaturing) Polyacrylamide Gel Electrophoresis (SDS-PAGE) of Proteins

Sodium dodecylsulfate (SDS) is a long-chain negatively charged detergent that binds to proteins. The hydrophobic portion of the detergent binds to the nonpolar regions of proteins, and the negatively charged sulfate group is exposed to water. When SDS binds to a protein, the protein noncovalent intra- and intermolecular associations (hydrophobic interactions, ionic bonds; hydrogen bonds) are disrupted. The protein treated with SDS is unfolded (denatured) to a random-coiled polypeptide. Since the surface of the SDS–protein complex is coated with negative charge from the sulfate groups of the SDS, the SDS–protein complex

---

[7] The ability to identify individual lactate dehydrogenase isozymes is of clinical importance for humans.

tends to form a rather rigid rod to minimize the interaction of the negative charges. The hydrodynamic volume of such a rod is approximately that of a sphere with a diameter of the rods length. Since the resulting negative charge on the SDS–protein complex vastly exceeds the limited number of intrinsic charges from the ionizable amino acid groups, all SDS–protein complexes have approximately the same charge-to-mass ratio and essentially the same shape. Such sos-complexes can be separated electrophoretically on the basis of size alone. The size of unknown proteins can be estimated by comparing their mobilities to those of proteins of known size (molecular weight standards). Since some proteins have intra- and/or intermolecular disulfide bonds, protein samples are usually heated in a boiling water bath in the presence of SDS and a high concentration of a thiol reagent [e.g., 2-mercaptoethanol (β-mercaptoethanol) or dithiothreitol (DTT)] before electrophoresis. The thiol reagents reduce intra- and interchain disulfide bonds to sulfhydryl groups. Those native proteins that are composed of several polypeptide chains (subunits) joined by intermolecular disulfide bonds will be transformed into their individual protein subunits. The sizes of these protein subunits can be estimated using SDS-PAGE (Fig 8.9), but the number of the protein subunits that combine to form the native protein cannot be determined. The SDS-PAGE electrophoretic procedure is very similar to that of native-PAGE. However, in SDS-PAGE all the buffers contain SDS in addition to the protein sample being treated with SDS.

SDS-PAGE is commonly used to estimate the number of different protein chains of different sizes contained in a protein preparation and to estimate the molecular weight of each of these chains. The number of different chains of different size is rather straightforward. The analyst just counts the number of protein bands in an SDS gel. The estimation of the molecular weight of the individual protein chains is based on a comparison of their mobility compared to that of protein chains of known molecular weight. The position of proteins (unknown and standards) is determined by protein staining. Then the relative

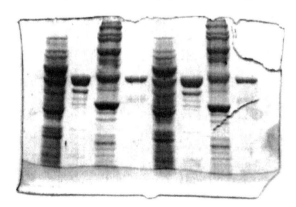

Fig. 8.9. A protein stained SDS-PAGE gel.

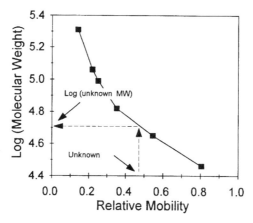

Fig. 8.10. The determination of a protein chain molecular weight from its mobility in SDS-PAGE.

mobility of each protein (unknown and standards) is calculated by dividing its migration distance of the protein from the top[8] of the gel by the migration distance of the tracking dye (see below) from the top of the gel:

$$R_m = \frac{\text{migration distance of protein}}{\text{migration distance of tracking dye}} \tag{8.4}$$

The relative mobilities of the proteins of known molecular weight (standards) are plotted (*x* axis) versus the logarithm of their molecular weights, log MW (*y* axis) (Fig. 8.10). Using this graph as a standard curve, the molecular weight of an unknown can be estimated from its relative mobility. Many such plots of relative mobility versus the logarithm of the molecular weight are linear. However, there can be significant deviation from linearity particularly at very high or very low molecular weights. Thus the analyst should use multiple molecular-weight standards, including those with molecular weights that are larger and smaller than the sample proteins. Molecular weights can be determined by interpolation between known standards if the plot is nonlinear. Estimates of protein molecular weights using this technique are accurate to about ±10%.

Sample Application

Sample solutions are usually prepared in a specified buffer in a high-density solution (frequently an inert concentrated sucrose or glycerol solution) and added

---

[8] For a discontinuous gel system, the migration distance is measured from the top of the separating gel to the position of the protein band.

to the sample wells of the gels just before the electrophoresis is begun. The high-density solutions are used to keep the sample from diffusing out of the well prior to the application of the electric field.

Applying the Electric Field and Running the Electrophoresis

All electrophoresis systems require a direct-current (DC) power supply and an electrode buffer system. The DC power supply provides the electrical potential for the electrophoresis, and the electrode buffer system maintains the pH and provides the ions that conduct the current across the electric field. In most cases the DC power supply is designed to provide a constant current at high voltages. There is always a tradeoff to be made in selecting the voltage to be used in electrophoresis systems. Higher voltages result in faster movement of the ions and thus faster separations. However, higher voltages also result in greater heat production from ohmic heating. The greater heat production can cause loss of biological activity, nonuniform voltages across the gels leading to differential movement of the analytes, and in extreme cases, melting of the gels. Many electrophoresis systems have built-in cooling systems. The reader is advised to follow the instructions for the individual electrophoresis apparatus as to the cooling and the voltages and currents to be used.

The voltages for the DC power supplies are often rather high (hundreds and sometimes thousands of volts), and thus there is a serious potential hazard for the user. There are several reports of serious injury and even deaths occurring when analysts have used electrophoresis systems without safety locks or with faulty safety locks. The reader is advised to use only those electrophoresis systems with good safety locks that are kept in proper repair.

Operationally it is usually important to determine the running time of the electrophoresis apparatus to ensure that proteins do not migrate off the bottom of the gel. For protein electrophoresis, the normal technique is to use a tracking dye (e.g., bromphenol blue) in electrode buffer and/or the sample buffer. At the high pHs normally used in protein electrophoresis, this dye has a negative charge and a blue color. The mobility of this dye is much greater than any protein, and the dye migrates at the interface of the electrode buffer and the gel buffer. Proteins migrate much slower. One can ensure that all proteins remain in the gel if the electrophoresis is stopped just before the blue dye band reaches the bottom of the gel. Thus the movement of the blue dye band allows one to track the progress of electrophoresis. The movement of the tracking dye is used in R_m computations.

Fixing

Once the electrophoresis process is finished, then it is necessary to visualize the biological molecules. The first step in the visualization of these molecules is to ensure that they do not diffuse out of the gel matrix before they can be observed and their travel measured. Except for those biological molecules with very high molecular weights (DNAs and RNAs), most biological macromolecules have sufficiently rapid diffusion that they will rather quickly (seconds to minutes) diffuse out of the gel during the staining process and be lost. Either the detection of the analyte must be done quickly or some technique must be used to prevent their diffusion out of the gel. Proteins are typically fixed in place using a common protein precipitating agent such as trichloracetic acid. Precipitation works well and is frequently used. However, precipitation of enzymes usually destroys their activity, and thus detection systems dependent upon their enzymatic activities cannot be used. Fixing and staining can be accomplished simultaneously (e.g., the Coomassie blue stain for proteins contains acetic acid and ethanol, which fix most proteins and the dye stains the proteins).

DETECTION OF PROTEINS AFTER POLYACRYLAMIDE GEL ELECTROPHORESIS

There are three general categories of protein detection techniques[9]: isotopic labeling, generic chemical reactions of all proteins, and the specific biological activity of a given protein (e.g., enzymatic activity) or its immunological properties (e.g., antigenicity).

Isotopic Labeling

The proteins in a gel can also be detected by labeling them prior to electrophoresis with a radioactive label such as ^{125}I, ^{32}P, ^{14}C, ^{3}H, and ^{35}S isotopes and then detecting the label by use of photographic film layered over the gel or some detection system that produces fluorescence in the presence of radioactive labels. Such detection systems are very sensitive but also have the usual hazards associated with radioactivity.

Generic Detection

The generic detection of proteins is usually done by staining, absorption of light at 280 nm, or fluorescence (rare). Most proteins contain amino acids that have aromatic residues that absorb UV light (~ 280 nm). It is possible to scan polyacrylamide gels using 280-nm light to locate the position of proteins within the gel. However, this technique can be technically difficult because of the background absorbance of the gel and the amount of aromatic amino acids varies from protein to protein. In addition, the specialized instrument required is quite expensive. Thus the absorbance of uv light is rarely used.

---

[9] The detection of DNA molecules will be discussed later in this chapter.

TABLE 8.2. **Common Protein Stains Used for Detection after Electrophoresis**

Stain	Advantages	Disadvantages
India ink	Cheap, sensitive	Enzyme detection often inhibited
Amido black	Cheap	Low sensitivity
Ponceau S	Cheap, compatible with antigen detection systems	Low sensitivity
Coomassie Brilliant Blue R	Sensitive	High background due to nonspecific binding
Silver stain—several versions	Very sensitive ∼100-fold more sensitive than Coomassie Brilliant Blue stains	Very high background due to nonspecific binding; enzyme detection often inhibited

There are several reagents that form colored complexes with the proteins, and these reagents can be used for the detection of the proteins in the electrophoretic gels (see Table 8.2). Most protein stains require destaining after the initial staining process.

Example: Coomassie Brilliant Blue Staining. A solution of Coomassie Brilliant Blue R in acetic acid (e.g., 10%) and alcohol (e.g., 25% ethanol) is commonly used to stain proteins in polyacrylamide gels. The gel is incubated in the solution for sufficient time (e.g., overnight at room temperature or 1 h at 50°C) to allow the Coomassie Blue to diffuse into the gel and bind to the proteins. After this "staining" reaction, the gel is "destained" to remove the Coomassie Blue from areas of the gel that do not contain protein. The "destaining" process involves incubation of the Coomassie Blue stained gel in an acetic acid[10]–alcohol solution (10%/25%). The destaining process is facilitated by changing the destaining solution or by including pieces of foam rubber in the incubation. The foam rubber adsorbs the Coomassie Blue as it diffuses out of the gel. The destaining process should be monitored, because the Coomassie Blue will eventually dissociate from the proteins. The presence of alcohol in this staining/destaining procedure causes the polyacrylamide gels to shrink. They can be easily swollen to their initial size by incubation in 10% acetic acid. The result of this process is the appearance of blue bands in the gels that are protein–Coomassie Blue complexes. Microgram levels of proteins can be detected in this manner. The stained gels can be photographed or/and dried between cellophane membranes to maintain a permanent record.

---

[10] The acetic acid and ethanol are used to do an acid precipitation of the proteins. Some low-molecular-weight proteins are not precipitated with this procedure.

Detection of a Protein Based on its Enzymatic Activity

It is frequently desirable to determine which, if any, of the protein bands seen in electrophoresis have the desired enzymatic activity and/or the desired immunological identity. Since the enzymatic activity of many proteins is retained after electrophoresis with native-PAGE systems, it should be possible to measure that enzymatic activity. However, almost all protein staining techniques destroy the inherent enzymatic activity when the protein is fixed in the electrophoresis gel. Thus alternate assay systems are needed to locate the specific position of the enzymatic activity within the polyacrylamide gel. A common technique is to chose a water-soluble substrate that will be transformed into a water-insoluble colored product by the enzyme of interest. After the electrophoresis is completed, the gel will be immersed in a solution of the substrate. At those locations in the gel that contain enzymatically active protein the soluble substrate will be transformed into an insoluble product that precipitates on the gel surface. The colored precipitate identifies the location of the enzymatically active protein. The enzyme activity staining process involves incubation of the gel in a solution containing the enzyme substrate(s) with a suitable buffer. The incubation continues until an adequate amount of the colored product has precipitated, and the enzyme reaction is stopped either by washing to remove the substrate or by the addition of a protein denaturation reagent (e.g., 5% trichloroacetic acid). Obviously, a rapid reaction is desirable, since the proteins can diffuse during the analysis of their enzymatic activities. Enzymatically determined electrophoretic protein bands often have less resolution than those where the proteins can be precipitated immediately after the electrophoresis.

"Suitable assays" have been developed for a variety of enzymes. For example, enzymes that catalyze the reduction of NAD^+ to NADH (e.g., alcohol dehydrogenase) can be coupled to the reduction of nitroblue tetrazolium to nitroblue formazan, a purple/blue insoluble compound (see Fig. 8.11). This coupled reaction is quite similar to that described for the spectrophotometric measure of dehydrogenase activity (see Chapter 5). Phenazine methosulfate (PMS) is required to facilitate electron transfer from NADH to nitroblue tetrazolium (Fig. 8.11). This assay is adaptable to any dehydrogenase by using the appropriate dehydrogenase substrate (e.g., ethanol for alcohol dehydrogenase, lactate for lactate dehydrogenase). Many other analogous assays have been developed for the location of enzymatically active protein bands.

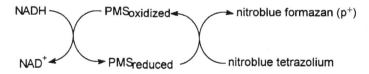

Fig. 8.11. The coupled reaction for the enzymatic detection of dehydrogenase enzymes in electrophoretic gels.

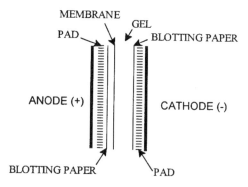

Fig. 8.12. The transfer system for Western blotting.

Detection of Protein Base on Antigenic (Immunologic) Properties (Western Blots)

As is discussed in detail in Chapter 9, mammalian species produce proteins called *antibodies* when they are challenged with foreign *antigens*. The technique called *Western blotting* (Western transfer) utilizes this specific interaction of antibodies with their antigen to locate and identify specific proteins in an electrophoretic gel. Since many of the antigenic activities of a specific protein are present after the proteins are fixed, Western blots are frequently used following slab SDS-PAGE. With Western blots, the electrophoresis is carried out in the normal fashion. At the end of the electrophoresis, the proteins are transferred with an electric field to a nitrocellulose[11] membrane to enhance the antigen–antibody reactions (see Fig. 8.12). Remaining binding sites on the nitrocellulose are then blocked using a protein that will not react with the antibody. The nitrocellulose-bound proteins are treated with a solution of antibodies (labeled or unlabeled). The antibodies will specifically bind only with their antigenic proteins. Finally, the labeled bound antibodies are detected using their radioactive or fluorescence label. Unlabeled bound antibodies can be detected using a second antibody in a manner similar to that described in Chapter 9 for the ELISA systems.

Isoelectric Focusing

Like electrophoresis, isoelectric focusing (IEF) is an electrokinetic technique that separates proteins on the basis on charge. However, in isoelectric focusing a pH gradient is established within the gel and a protein will move to a pH region where it has no net charge. The pH gradient in isoelectric focusing is established by using ampholytes that are mixtures of polymers of weak acids and bases. The anode (+) reservoir contains a dilute acid solution, and the cathode (−) reservoir contains a dilute base solution. When an electric field is applied, the ampholytes,

---

[11] New plastics and nitrocellulose have a rather high capacity to bind all proteins.

which are also the major source of ions in the system, migrate toward the anode or cathode according to their charge distribution. For example, ampholytes with a preponderance of weak acid groups (e.g., carboxylic acids) move toward the anode, while those with mostly weak base groups (e.g., amines) migrate toward the cathode. This movement of ampholytes creates the pH gradient with lower pH values near the anode (+) and higher pH values near the cathode (−).

A protein placed in this gradient will migrate according to its net charge at the pH of its local environment. If the pH of the local environment is above the pI[12] of the protein, the protein will have a net negative charge. The net negative charge will result in movement of the protein toward the anode (+) and into a lower pH environment. At lower pH, the net negative charge on the protein will be reduced. Eventually, the protein will encounter a pH environment at which it has no net charge, the isoelectric point (pI). At this point, there is no driving force on the protein from the electric field. The protein will form a sharp, focused band at this point in the gradient because diffusion in either direction moves the protein into a pH environment in which it will have a net charge. For example, if the protein initially located at pH = pI diffuses toward the anode, it moves into a pH environment where pH < pI. Under these conditions, the protein will have a net positive charge and will be pulled toward the cathode (−), a higher pH environment, where it loses its net positive charge. If the protein diffuses into a higher pH environment, pH > pI, it will have a net negative charge and will be pulled toward the anode (+). The pI of the protein can be determined by its location within the pH gradient in the isoelectric focusing gel. Isoelectric focusing has a high potential resolving power. Since isoelectric focusing only uses charge to separate proteins, it is a nice complement to native PAGE, which utilizes both charge and size to separate proteins.

Two-Dimensional Electrokinetic Systems

Two-dimensional electrokinetic systems are a combination of isoelectric focusing and SDS-PAGE. Proteins are separated based on their charge characteristics with isoelectric focusing in a tube gel. The gel from isoelectric focusing is placed on top of an SDS-PAGE stab gel. Electrophoresis within the SDS-PAGE system separates the proteins based on their sizes. Thus these proteins have been separated by charge in one dimension and by size in a second dimension oriented 90° from the first. Two-dimensional electrokinetic systems have very high resolving power. There are reports of separation of mixtures of over a thousand different proteins. On the other hand, two-dimensional electrokinetic systems are technically more difficult to use than the more traditional one-dimensional electrophoresis systems.

---

[12] The pI of a protein is that pH at which the net charge on the protein is zero.

Agarose Gel Electrophoresis of DNA

While protein molecules have molecular weights from about 13,000 to about 350,000 D, DNA molecules are much bigger, with molecular weights typically in the millions. While small DNA molecules (<1000 base pairs,[13] bp) can be separated using large-pore polyacrylamide gels (e.g., for DNA sequencing), the pore sizes of usable polyacrylamide gels are too small for separation of the larger DNA molecules. The polyacrylamide gels with very large pores suitable for separation of DNA molecules are too fragile to be used analytically. Gels prepared from agarose have larger pore sizes and adequate stability. Thus agarose gels are usually used for electrophoretic separation of DNA molecules.

Agarose is an unbranched linear polysaccharide. Agarose dissolves in aqueous solution upon heating and when cooled forms a solid gel consisting of left-handed helices that intertwine and have pore sizes of the proper size to accommodate large DNA molecules. The agarose gel pore size can be controlled by varying the concentration of agarose in the solution used to prepare the gel. As the concentration of agarose increases, the pore size decreases. The agarose concentration can be varied from 0.5 to 8% (w/v). One or 1.5% agarose gels are probably the most frequently encountered.

Agarose gels are poured in a horizontal orientation. A comb is placed in the agarose solution before it cools to form the solid gel. The comb forms wells to which the DNA samples are added in much the same manner as a comb is used to form wells in polyacrylamide gels for protein electrophoresis. The negatively charged phosphate in every base pair gives every DNA molecule a large negative charge and a rather uniform charge-to-mass ratio. Thus, in an electric field, DNA molecules will move toward the anode (+). The separation of DNA molecules in the pores of the agarose gel is based on size, since all DNAs have a constant charge-to-mass ratio. DNA has no significant absorption of light in the visible region, and thus other means must be used to visualize the location of DNA molecules within an agarose gel. Common techniques for visualization of the DNA are isotopic labeling (typically with P^{32}) or by staining ethidium bromide[14] (see Fig. 8.13). Since DNAs have no visible color, tracking dyes (e.g., bromphenol blue) are frequently used to monitor the electrophoretic process. Note that some tracking dyes do not move faster than small nucleic acid molecules,[15] and thus mobility rather than relative mobility is used for nucleic acid agarose electrophoresis.

---

[13] The average molecular weight of a DNA base pair is about 660 D.

[14] Ethidium bromide intercalates into the DNA structure, and the ethidium bromide–DNA complexes fluoresce orange when exposed to ultraviolet light. Ethidium bromide is a carcinogen, and appropriate care must be taken when it is used to stain DNA.

[15] Note the contrast to protein PAGE.

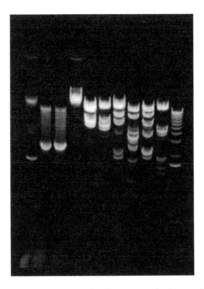

Fig. 8.13.   An ethidium bromide stained agarose gel of several DNA samples.

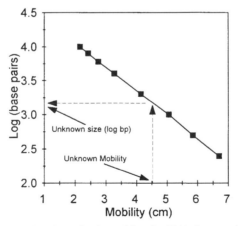

Fig. 8.14.   The determination the molecular weight of a DNA fragment from its mobility in an agarose gel. Note, the similarity to the system for determining the molecular weights of protein chains in SDS-PAGE.

Since the separation of various DNAs is based upon their size and not their charge, the molecular size of DNA molecules can be determined using a method similar to that used to determine the size of an unknown protein after SDS-PAGE (see Fig. 8.14). The agarose electrophoresis system is calibrated by concurrently running DNA standards of known molecular weight and comparing the mobility of the known standards with that of the DNA molecules of unknown size. With DNA agarose electrophoresis, the migration distances (not relative mobility, R_m) are plotted (x axis) versus the logarithm of their sizes. Since DNA sizes are frequently reported as number of base pairs per molecule (#bp), the y axis has units of log(#bp). A typical standard curve for the estimation of the size of a given DNA species is shown in Figure 8.14. Not all such plots are linear.

Not only is DNA agarose electrophoresis used to determine the molecular weight of various DNAs, this powerful tool is also used in DNA sequencing. There are several excellent discussions on the sequencing of DNA, and the reader is referred to a good biochemistry or molecular biology textbook for details.

9.   Enzyme-Linked Immunosorbant Assays

BACKGROUND AND OVERVIEW

In mammals adaptive immunity is conferred upon the animal by cells called *lymphocytes*. The lymphocytes synthesize proteins called *antibodies* when the animal is challenged by foreign molecules (antigens). The antigens are tightly bound to these antibodies at the antigen recognition sites (see Fig. 9.1). Each lymphocyte cell can recognize only one antigen, and within any given animal there are approximately 10 million different binding sites for antigens. The selectivity of the specific lymphocyte proteins in the binding of the antigens can be very high. In some cases it is the most selective binding known between two given compounds. Even when there are different antibodies to an antigen, the normal situation is that each different antibody recognizes and is bound to different structural components of the antigen.

The particular structural feature of the antigen that binds to the antibody is called an *epitope*. If the antibodies are derived from a single cell line (*monoclonal antibodies*), then all the antibodies will recognize the same epitope. If the antibodies are derived from a series of different cell lines (*polyclonal antibodies*), the antibodies will recognize as many epitopes as there are cell lines. Monoclonal and polycolonal antibodies are used for chemical analyses in biological systems.

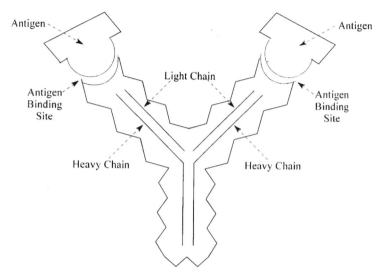

Fig. 9.1.   Diagram showing IgG structure.

Monoclonal antibodies have the highest selectivity but are more expensive to produce. To produce monoclonal antibodies, the lymphocytes must be isolated and grown in pure culture, (monoclonal culture). Most cell lines go through a limited number of cell divisions before the lines die. Systems have been developed for fusing antibody-producing cells with *myeloma* cells (a type of cancer cells that grow well in cell culture and are generally viewed as immortal). The resulting hybrid cells are called *hybridomas* (hybrid myelomas). They can be readily grown in cell culture and permit the production of large quantities of selected antibodies.

Polyclonal antibodies are frequently used in sandwich enzyme-linked immunosorbant assays (ELISA) (see below) and are less expensive to produce. The polyclonal antibodies are usually isolated from blood sera. With today's technology one can make large quantities of antibodies against almost any compound. What is needed is a skilled technician, six months to a year, and adequate funding. More recently the techniques of biotechnology are being used for the production of monoclonal antibodies.

There are several different antibody types (IgA, IgG, IgM, etc.), but for simplicity only IgG (immunoglobulin G) will be considered here. IgG molecules are proteins composed of four subunits (two smaller polypeptides called light chains and two larger polypeptides called heavy chains). These four polypeptides combine to form a Y-shaped molecule (Fig. 9.1). The tips of the Y contain the antigen-binding (antigen-recognition) sites. The two antigen-recognition sites on an IgG molecule are identical such that a single IgG molecule can bind to two different antigen molecules with identical epitopes.

The binding of multiple antibody molecules to a single protein (antigen) allows the development of a variety of methods, some of which can be adapted to small molecules that have only a single epitope. For example, if antibodies cannot be directly made to some low-molecular-weight analyte,[1] they often can be made indirectly to the same molecule by the following strategy. First covalently bind the analyte to a medium-size protein such as bovine serum albumin. Second, prepare antibodies to the analyte–bovine albumin conjugate. Multiple antibodies will be made, some to the analyte, some to the bovine serum albumin, and some to the analyte–bovine albumin conjugate. Third, prepare an affinity chromatography column using bovine albumin as the active binding site. Fourth, chromatograph the antibodies prepared against the analyte–bovine albumin conjugate on the bovine serum albumin column. The antibodies whose epitopes include any portion of the bovine serum albumin will be retarded on the column; those antibodies that will bind only to the analyte epitopes will elute first from the column. Thus the analyst now has antibodies for ELISA assays of the analyte.

Definition: Antigen. The compound that elicits the antibody to response in an animal.

Definition: Antibody. The protein whose concentration is amplified in the animal in response to the antigen challenge. One antibody molecule can bind two antigens with identical epitopes.

Definition: Epitope. The particular structural feature of the antigen that is recognized by and binds to the antigen recognition site of the antibody.

Definition: Monoclonal Antibodies. Antibodies from a single purified cell line. These antibodies have a recognition site for a single epitope. Only one chemically identifiable antibody is produced by each cell line.

Definition: Polyclonal Antibodies. Antibodies from multiple cell lines. There will be as many different kinds of antibodies molecules as there are cell lines. Thus polyclonal antibodies recognize multiple epitopes on the antigen.

The terminology for naming antibodies and antigens can be a bit daunting, and thus it is worthwhile to present some of the simpler concepts. If one uses the IgG proteins[2] from a rabbit as antigens in the production of antibodies in goats, then

---

[1] The low-molecular-weight analytes that elicit specific antibodies are sometimes called *haptenes.*

[2] IgG is a representation for a class of proteins. There are many different proteins in this classification.

the goat will produce antibodies[3] named "goat antirabbit IgG." These proteins named "goat antirabbit IgG" would be goat IgG proteins. One might then take the goat IgG proteins and use them as antigens to produce antibodies in rabbit. Then one would have "rabbit antigoat IgG" antibodies.

The availability of pure antibodies has led to an immunotechnology for identifying and quantifying of complex molecules: the immunoassays. There are a large variety of immunoassays. This book will concentrate on the immunosorbent assays that take advantage of the fact that proteins will quantitatively bind to nitrocellulose and new plastic, especially to polystyrene and polyvinyl chloride. This binding is normally not reversible and is inhibited by hydrophobic substances such as oil, other proteins, and other hydrophobic compounds. With ELISA assays the antigens (which are often proteins) or the antibodies (which are always proteins) are bound (adsorbed) to the solid surface of 96-well microtiter plates or 8-well microtiter strips, which are made from either polyvinylchloride or polystyrene.

Most antibodies are not easily visualized by themselves, and it is difficult to directly quantify their levels. Thus reporter group technologies have been developed to quantify antibodies. The concept is that the specific antibody or antigen is labeled so that it can be readily measured. Initially the preferred technique was to covalently link a radioactive molecule to the antibody or antigen and then measure radioactivity. Such assays, called *radioimmune assays*, are selective and sensitive. Almost any radiochemistry detection system can be used for the measurement of the bound labeled antibody or antigen. The assays can be used in a direct measurement mode or in a competitive mode. However, there are significant problems related to the safety of their use and the disposal of the radioactive materials at the end of the assays. Most radioimmune assay uses have been restricted to research situations, clinical laboratories and specialized analytical laboratories. Many laboratories that once made extensive use of radioimmune assays are now turning to ELISA assays.

Fluorescent molecules such as fluorescein have also been used as reporter groups. In this case a fluorescent molecule is covalently linked to the antibody. The amount of antibody is then determined by measuring the fluorescence of the system. Such assays work well and are used extensively in cell staining. Their use in quantitative assay is more limited due to their relatively low sensitivity and their requirement for rather expensive instrumentation.

Enzymes have become a popular reporter group technique used for the quantification of the level of specific antibodies or antigens. Enzyme activity is not normally a natural component of an antibody. An enzyme must be covalently attached in vitro to the antibody or the antigen using chemical cross-linking reagents. The products of these in vitro reactions are antibody–enzyme or antigen–enzyme conjugates. Alkaline phosphatase or horseradish peroxidase are

---

[3] Antigens of the size of IgG proteins have multiple cpitopes and will elicit responses leading to the production of multiple antibodies (polyclonal antibodies).

the reporter enzymes most frequently used because of their low cost, their stability, and the availability of substrates that yield water-soluble colored or fluorescent or chemiluminescent products. The amount of the enzyme–antibody conjugate is determined by measuring the amount of the enzyme activity present.[4] Since one enzyme molecule can produce many product molecules, this system is frequently called an *enzyme amplification system*.

The immunosorbent[5] versions of the enzyme amplification assays are called enzyme-linked immunosorbent assays. ELISA assays[6] are some of the most sensitive and selective assays available for modern analyses. The assays obtain their selectivity from the selectivity of the antibody or antigen and their sensitivity from the use of enzymes as reporter groups. ELISA assays have become the method of choice for making quantitative immunoassays. This book will limit its discussion of immunoassays to a discussion of ELISA assays, and the reader is directed to other appropriate references for discussions of the other immuno-assays.

ELISA CLASSES AND VARIATIONS

ELISA methods can be subdivided into two types (1) two antibody assays (see Fig. 9.2) and (2) antibody capture (see Fig. 9.3). The two antibody assays will be discussed first, since they appear to be the method of choice in most laboratories.

Two-antibody assays are commonly called "sandwich assays," and they are used to determine the concentration of known antigens. In this assay, the antigen is "sandwiched" between two antibody molecules. This procedure requires affinity-purified polyclonal antibodies (antibodies that bind only to the antigen of interest) or two monoclonal antibodies. The monoclonal or polyclonal antibodies must bind to different nonoverlapping epitopes on the antigen. Unmodified and modified (antibody–enzyme conjugate) antibodies are used. The steps in the sandwich assay are:

1. Obtain an unlabeled antibody to the analyte
2. Add and bind the unlabeled antibody to the analyte to wells of the microtiter plate

---

[4] Obviously, for the enzyme to be an effective reporter group, it must retain its enzyme activity after being covalently bound to the antibody or antigen. In many cases the enzyme activity is retained, although it is frequently reduced in the binding process. Likewise, the antibody must retain its antigen recognition site after being covalently modified. Thus the activity of any enzyme–antibody conjugate must be calibrated prior to its use in an assay.

[5] These antibody–enzyme conjugates are also used in Western blotting (see Chapter 8).

[6] The authors are aware that there is a redundancy in saying ELISA assays. However, that is the standard accepted terminology.

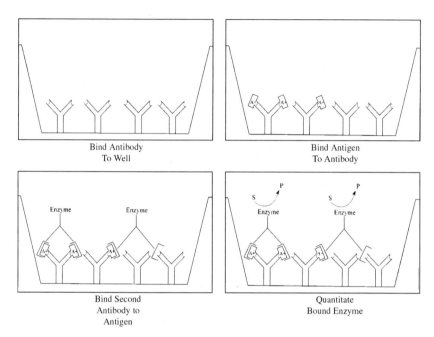

Fig. 9.2. Cartoon showing step-by-step procedure of two anti-body sandwich (ELISA).

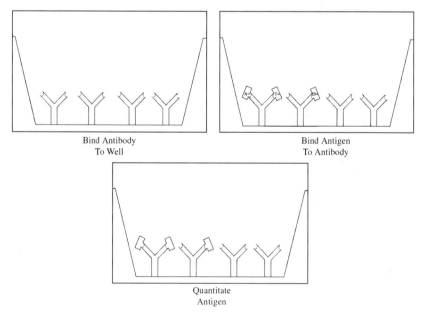

Fig. 9.3. Cartoon showing step-by-step procedure of antibody capture assay.

3. Wash wells to remove unbound antibody

4. Add a blocking protein[7] solution to wells to block the remaining available protein binding sites on well surfaces

5. Wash to remove unbound blocking protein

6. Add and bind of analyte[8] (antigen) to wells

7. Wash to remove unbound antigen

8. Add and bind enzyme-labeled antibody to analyte

9. Wash to remove unbound enzyme-labeled antibody

10. Add enzyme substrate to start enzyme reaction

11. Stop the enzyme reaction[9]

12. Qualify product of enzyme-catalyzed reaction using an ELISA plate reader

The binding steps (2, 4, 6 and 8) are diffusion controlled and require a significant waiting period, typically 2 or more hours at room temperature. Such binding steps are frequently allowed to proceed overnight at 4°C.

Different lots of microtiter plates have different antibody binding capacities. Furthermore, the binding to the titer plate is influenced by the chemical composition of the entire assay solution, and thus changes in that chemical composition will change the binding to a given titer plate. It is always necessary to determine the optimal concentration for the initial binding to the titer plate on a case-by-case basis. Furthermore, different lots of enzyme-conjugated antibody may have different binding capacities for the antigen of interest. It is always necessary to determine the optimal concentrations for the antigen–antibody binding on a case-by-case basis. Thus, when running ELISA assays, a wise analyst analyzes known concentrations of the analyte for the development of standard curves over a wide concentration range so that the usable linear portion[10] of the analyses can be found in any given assay. Likewise, a wise analyst will analyze the sample with an unknown concentration of analyte over a series of concentrations. Table 9.1 shows typical concentrations for the analyses of the known and unknown concentrations of the analyte. Typical response curves are shown in Figures 9.4 and 9.5. Note that there is a significant portion of these response curves where the slope of the line is essentially flat due to saturation of the bound antibody–antigen binding sites. Note also that there are portions of the response curves that are above

[7] Usually bovine scrum albumin, BSA.

[8] Standard curves and the samples of unknown analyte concentration are usually run in the same batch. See below.

[9] Note that the sensitivity of the assay can be changed by altering the time period allotted for the enzyme reaction.

[10] While these portions of the binding curves may not be truly linear, they approach linearity at low analyte concentrations.

TABLE 9.1. **A Typical Dilution Table for an ELISA Assay;**
Rabbit IgG is the Analyte, and its Concentration is being
measured in Rabbit Sera

Column	[Rabbit IgG] (μg/mL) 0.1 mL/well Rows A and B	[Rabbit serum] (μg/mL) 0.1 mL/well Rows C and D
1	0	0.025
2	0.005	0.050
3	0.01	0.075
4	0.02	0.10
5	0.03	0.25
6	0.05	0.50
7	0.10	0.75
8	0.20	1.0
9	0.30	2.5
10	0.50	5.0
11	1.0	7.5
12	2.0	10.0

the linear range of the assay. With both the standard curves and the response curve for the analysis of the unknowns it is usually necessary to limit the presentation to the linear range portion of the standard curve (see Figs. 9.6 and 9.7)

Calculations

Once the linear portion of the standard and unknown response curves has been established, the computation of the concentration of the analyte is straightforward:

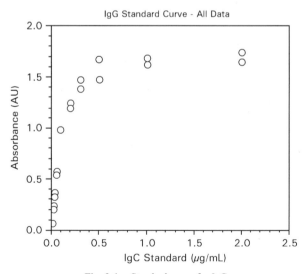

Fig. 9.4.   Standard curve for IgG.

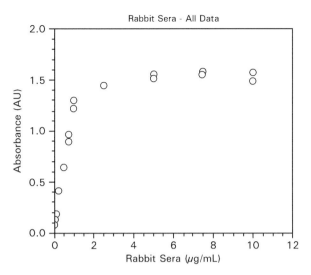

Fig. 9.5. Response curve for IgG analyses of rabbit sera.

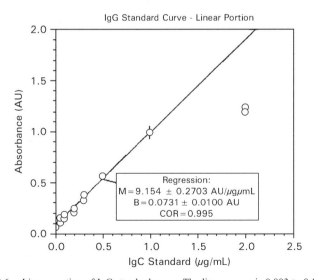

Fig. 9.6. Linear portion of IgG standard curve. The linear range is 0.003 to 0.1 μg/mL.

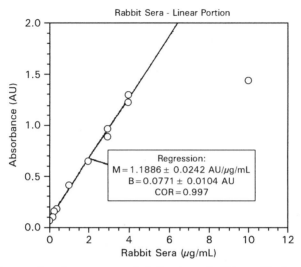

Fig. 9.7. Linear portion of response curve for IgG analyses of rabbit sera. The linear range is 0.025 to 4.0 μg/mL.

concentration of the analyte in the unknown

$$= \frac{\text{sensitivity of response curve of unknown}}{\text{sensitivity of standard curve}} \quad (9.1)$$

Example: For the data shown in Figures 9.6 and 9.7:

$$\text{concentration}_{\text{IgG}} = \frac{\dfrac{\text{AU}}{\mu g_{\text{rabbit sera}}}}{\dfrac{\text{AU}}{\mu g_{\text{IgG}}}} = \frac{\mu g_{\text{IgG}}}{\mu g_{\text{rabbit sera}}}$$

$$= 0.13 \text{ mg IgG/mg rabbit sera}$$

Antibody Capture Assays

Antibody capture assays (Fig. 9.3) are quite similar to sandwich assays. The difference is that with antibody capture assays the protein antigen is bound directly to the microtiter plate wells. The bound antigen is either directly determined by binding an enzyme-conjugated antibody followed by a measurement of the enzyme activity or indirectly determined by the binding of an antibody to the analyte (antigen) followed by the binding of an enzyme-conjugated antibody prepared against the first antibody followed by a measurement of the enzyme activity. The advantage of using a labeled second antibody reagent is that many are commercially available, thus eliminating the need to produce a different antibody–enzyme conjugate for each antibody capture ELISA method. For example, if the objective is to determine whether a human blood sample contains antibodies against a particular microorganism, an antihuman immunoglobulin antibody–enzyme conjugate could be used. The antigen(s) (microorganism proteins) would be bound to the polystyrene, the human serum to be tested for the antibodies against the microorganism would be added followed by the antihuman immunoglobulin antibody–enzyme conjugate. The conversion of enzyme substrate to colored product provides evidence that the antibody was present in the blood sample. The antibody capture assay methods require pure antigen for system calibration, but the primary antibody need not be in purified form. Thus the presence of a specific antibody in a serum sample can be assessed using a pure antigen and a labeled commercially available secondary reagent. The amount of colored product formed is a direct measure of the amount of antibody present in the sample. These graphical data obtained are comparable to those obtained in the sandwich ELISA procedure.

CAUTIONS ON THE USES OF ELISA ASSAYS

Accuracy

Unfortunately, many antibodies cross-react with antigens that are not the analytes of interest but are closely related to the compounds of interest. Studies have shown that there is a great deal of immunological similarity between the different compounds found in biological systems. This is particularly so within families of proteins and complex carbohydrates. While these similarities are useful in establishing genetic genealogies, etc., the same similarities can create potential sources of error when using antibodies to identify and quantify analytes. For example, it is known that many antibodies to the proteins and complex carbohydrates of pathogenic *Salmonella* also react with the proteins and complex carbohydrates of nonpathogenic *Salmonella*. Thus if an analyst were using antibodies to pathogenic *Salmonella* to determine the concentration of pathogenic *Salmonella,* there would be a serious potential for the presence of nonpathogenic *Salmonella* to give a positive response even if there were no pathogenic *Salmonella* present. Thus when an analyst prepares to use an ELISA assay for the determination of an analyte in a specific matrix, it is usually necessary to demonstrate that there is not cross-reactivity of the assay antibody with the non-analytes normally found in the matrix.

Precision

As was mentioned earlier, different lots of microtiter plates have different binding capacities for the antibodies of interest, and different lots of enzyme conjugated antibody also have different binding capacities for the antigen or antibody of interest. Furthermore, it is known that some components of biological matrices interfere with the binding of antibodies with antigens. Given these potentials for variability, it should not be surprising that the results obtained from ELISA assays usually have quite high relative standard deviations. ELISA analyses are frequently run in triplicate so that the relative standard deviations are within acceptable limits. Even with triplicate analyses acceptable correlation coefficients (R^2) for ELISA assay are frequently lower than those for the more traditional assays. Thus while an analyst might set a standard of acceptance of $R^2 \geq 0.99$ for most analyses, the same analyst will frequently accept an $R^2 \geq 0.95$ for an ELISA analysis.

GENERAL REFERENCE

Ed Harlow and David Lane, "Antibodies—A Laboratory Manual," Cold Spring Harbor, 1988.

10.   Assay Quality Control and Data Validation

INTRODUCTION

A common tendency is to consider any chemically determined value a "magic number," that is, the "true" value. This is a dangerous misconception. As any working analyst knows all too well, analytical values are all too frequently incorrect. Mistakes can be made by the analyst in the selection of the samples, in the choice of methods, in the performance of the method, in the recording of the data, in the computation of the final results, in the reporting of the results, etc. Other mistakes are due to the failure of instrumentation, the failure of reagents, the failure of computing programs, etc. The complexity of biological samples and the low levels of many metabolites lead to frequent interferences with the analyses of individual analytes in biological samples. Since anyone can make mistakes, assay quality control is essential if the analyst is to obtain meaningful results of acceptable accuracy and precision. The three goals of analytical assay quality control are to minimize the variance of the measurements, to verify the accuracy of the measurements, and to document the precision and accuracy of the measurements (see Fig. 10.1). The use of good analytical quality assurance systems gives credibility (believability) to the analytical data and provides the

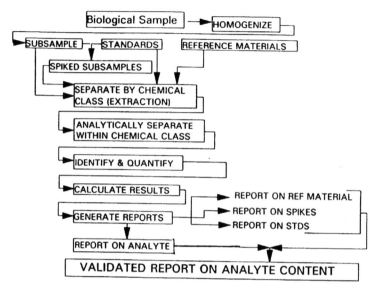

Fig. 10.1. An idealized assay method. Reprinted (slightly modified) with permission from *Nutrient Analysis of Foods: State of the Art for Routine Analysis* (1979) Kent K. Stewart (Ed.) Copyright, 1979 by AOAC INTERNATIONAL.

TABLE 10.1.　Elements of Good Assay Quality Control

Documentation
    Notebooks, SOPs, methods, materials, solutions, instruments, samples, software, etc.
Standard methods
    Written SOPs
    Published
    Certified
Instruments
    Calibration
    Instrument logs
    Preventive maintenance
Controls
    Internal standards (especially for chromatography)
    External standards
    Method of standard additions
Reference samples
    In-house pool samples
    Commercial reference materials
    Standard reference materials (NIST)
Certified algorithms and computer programs
    Cross-check computation algorithms
    Computer hard disk backups
Audit trails
    Be able to follow all samples completely through all assays
Assay quality-control officer

analyst with a means of identifying unreliable data and problems in an analytical process. The elements of good analysis quality control are given in Table 10.1.

DOCUMENTATION

Documentation is the key to good assay quality assurance. The analyst needs to ensure that adequate bench notebooks are kept and that all the laboratory procedures have written standard operating procedure (SOPs[1]). Concise description of the methods should be provided with citations of literature references or published methods on which the method is based. Instrument and balance logs should be kept, and the routine calibrations and preventive maintenance should be recorded in them. All chemicals should be logged into the laboratory inventory with an indication of the date of receipt, lot number, and source. All preparations of solutions and reagents should be recorded in the bench book with an indication of who prepared the solution, which SOP was used, and which chemical lots were used. All chemicals and preparations should have expiration dates marked on their containers. The individual reagent containers should be labeled with the date of preparation and a reference to the preparation log. The refrigeration and freezer storage facilities should have a temperature calibration, and the daily temperatures should be logged. All programs and software should be identified and a copy should be archived. All computer systems should be backed up at least weekly, and the backup disks or tapes should be stored off-site.[2] Audit trails should be established so that an analyst is able to follow all samples completely through all analyses with written documentation (see Fig. 10.2). Final report forms, assay numbering, assay logs, data filing systems, and other data reporting/tracking requirements are also needed. These may seem like excessive documentation requirements, but the experience of the authors is that failure to keep such documentation will sooner or later result in a disaster.

METHOD VALIDATION

An analyst usually has a variety of reasons for picking an assay method for a particular analyte in a specific biological matrix (see Chapter 11). However, after an assay methodology has been selected, it should be specifically validated for the

---

[1] SOPs (standard operating procedures) are written step-by-step descriptions of the entire method as it is actually preformed. Most modern analytical methods are much too complex to be done from memory. A good SOP will have been prepared by one analyst and checked by another. Any deviations from SOPs must be documented in the bench notebook.

[2] Computer hard disks do crash; it is not a matter of if, but of when.

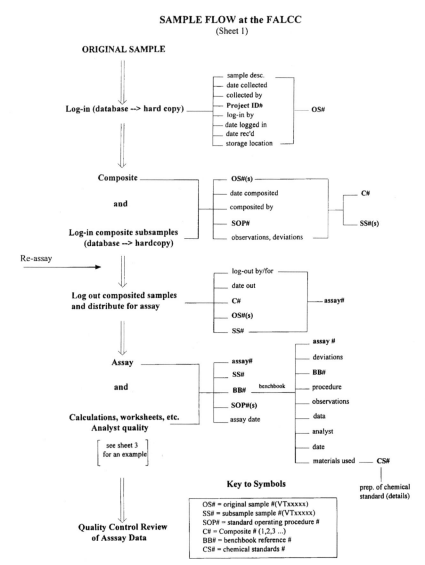

Fig. 10.2. A rather complex and detailed audit trail. Private communication from K. M. Phillips, FALCC, Department of Biochemistry, Virginia Polytechnic University and State University. Reprinted with permission. Continues on next page.

analyses of desired analytes in the current sample matrices. Such validations will alert the analyst to possible problems due to matrix effects, and successful validations will give credibility to the analytical results. A detailed discussion of method validation is given in Chapter 11.

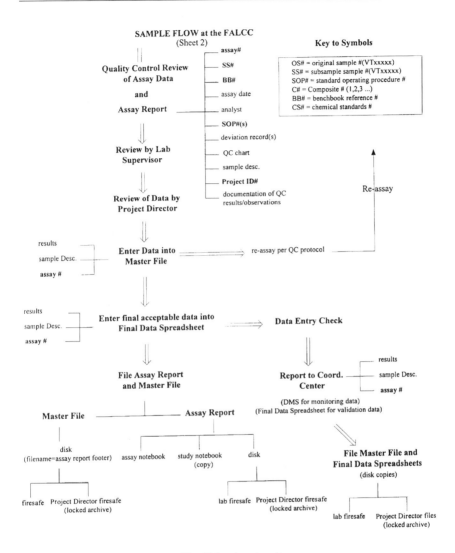

Fig. 10.2. (*continued*)

VALIDATION OF DATA SETS

Method validation is not sufficient to produce believable data. Any analyst can make a mistake even when using a validated method. Given the potential for errors in the chemical analyses of biological samples, the authors believe that it is prudent for the analyst to validate each individual analytical data set (Table 10.2). Such data validation can save the analyst many painful and embarrassing

Table 10.2.   Data Set Validation

Replicates
Pure standards
Sample blanks
Pool samples
Reference materials
Method of standard additions
Quality-control charts

moments. The time spent in developing and utilizing such procedures will usually provide ample payback. Validation of individual analyses always starts with adequate description of the samples and documentation of the regents and methods used. Almost always the assay system will require a calibration back to a known concentration in a sample of known analyte composition. Analysts need to report the figures of merit in their assays, what blanks were used, what standards and positive controls were used, and how the computations were made.

REPLICATE SAMPLES

At the minimum the analyst should analyze each sample in duplicate. This replicate analysis is not done for any statistical reason; it is done to reduce the probability of random disasters in which the individual analysis just went wrong. The concept is that truly random errors will not be repeated in the short term. Two analyses of the same sample that give the same result give the analyst some confidence in the results. Basically one should never (well, almost never) report any important result based upon one analysis. Most analytical problems require more than simple replicate analyses for data validation.

ANALYSES OF PURE STANDARDS AND STANDARD REFERENCE MATERIALS

One of the more powerful means to validate data is the analysis of samples of known concentrations in the same batch with the unknowns to demonstrate that the assay performed as expected. Typically an analyst will include a set of pure standards and a sample blank with each set or batch of assays. For even better data validation the analyst will require the analysis of blanks, standards, pool samples, and/or reference materials[3] (blind-coded) with each batch of samples or

---

[3] Primary Standard Reference Materials (SRMs) are homogenous analytical materials whose analyte concentrations have been determined by two independent analytical methods. They are prepared by the National Institute of Standards and Technology. Secondary reference materials whose composition can be traced back to the primary SRMs are frequently used to reduce the cost of analyses, since SRMs are relatively expensive.

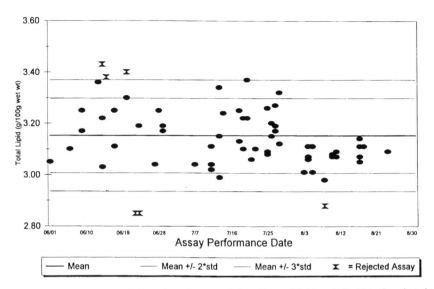

Fig. 10.3. A quality-control chart. A quality-control chart for total lipid analysis. Note the rejected assay lots. The method appears to have drifted from 8/3 though 8/21. Private communication from K. M. Phillips, FALCC, Department of Biochemistry, Virginia Polytechnic University and State University. Reprinted with permission.

at routine intervals say every fifteenth sample. Each of these standards fulfills a different purpose. The analysis of pure standards ensures that any change in the response of the assay system will be seen. The zero sample blank provides a check for baseline drift. The analysis of pool samples and/or the reference materials provide a means for checking for method drift. Careful workers will often include a series of spiked samples[4] with different concentrations of added analyte to check that the total response factor is not changing.

QUALITY-CONTROL CHARTS

The results of the analyses of the standards and quality-control materials should be plotted on a quality-control chart such as shown in Figure 10.3. Quality-control charts are used to demonstrate that the entire analytical process is in control (or not). Typically the mean and standard deviation of the assay methodology is established by analyzing a series (about 15) of standards and quality-control materials over several days. It is best for these measurements to

---

[4]Spiked samples are those where a known quantity of analyte was added to the sample matrix. See Chapter 11 for a more detailed discussion of spiked samples.

span the most probable sources of variance (different analysts, time, new batches of reagents, etc.) expected across the duration of the study. These data are used to establish the mean and confidence limits for the assay. The results of each analysis of the standards and quality-control materials are plotted as shown in Figure 10.3. An assay methodology that is behaving properly will have these results randomly distributed about the mean, with about 95% of the values within plus or minus two standard deviations of the mean. In those cases where the results of the analyses of either the standards or the quality-control materials are outside the mean plus or minus three standard deviations, the analyses are said to be out-of-compliance, and all the data from that batch of samples should be discarded and the samples reanalyzed. If reassay yields a control value out-of-range, further analyses should be stopped until the reason for the lack of compliance is investigated and corrected.

Sometimes assay systems will drift slightly up and down over time for no identifiable reason, and this can be considered part of overall analytical variance. The quality control charts are also used to check for significant method drift. If the results of the analysts of the quality control materials for seven or more batches of samples are all more than two standard deviations away from the mean, on the same side of the mean, then the method is probably drifting. While such drifting may be *statistically* significant, its practical implications will be determined by the magnitude of the drift and its impact on the use of the data in a particular study. No general methods have been published yet for determining acceptable limits for assay drift based on end use of the data.

REVIEW OF QUALITY ASSURANCE AND DATA VALIDATION

While it is known that errors in the chemical analysis of biological systems are common, it can be difficult to determine when such errors occur. One might think that the analysts doing the analyses would be the best people to catch errors. Indeed the analysts are the first line of defense. However, analysts are almost always personally involved in the results of their analyses. Given human nature, asking analysts to review their own work is unlikely to elicit unbiased views of the results. Thus it is wise to have external review of the assay quality-assurance systems and data set validations. Outside review can be as informal as having a colleague review the procedure for assay quality control and/or the data validation systems for a data set. Many errors, omissions, and mistakes will be caught by such an informal review. Laboratories with heavy analytical loads often have one or more quality-control officers whose job it is to review all aspects of assay quality-assurance systems and data set validations. Ideally such quality-control officers are not supervised by the analysts doing the analyses but rather are of equal rank to the analyst and report independently to a common supervisor. Many laboratories have their quality-control systems reviewed by people or organizations outside the laboratory. Sometimes such reviewing groups are legally constituted or have a semilegal status. The greater the impact of the analytical data on commerce or human health, the more likely the review process will have a legal or semilegal status. The peer review system for scientific publication of data and for the awarding of financial support of research are other types of external review.

IMPACT OF QUALITY-CONTROL SYSTEMS

Assay quality control is done so that the analyst can provide analytical data of acceptable accuracy and precision. The use of good analytical quality-assurance systems gives credibility (believability) to the analytical data and provides the analyst with a means of identifying unreliable data and problems in an analytical process. While use of quality-control samples provides documentation of assay performance, it is incumbent upon the investigator to interpret the significance of any deviations in the context in which the sample data are used. While statisticians can offer advice in such cases, the evaluation of the impact is usually not a statistical question.

SUGGESTED READING

Taylor, J. K. *Quality Assurance of Chemical Measurements.* Lewis Publishers, Inc., Chelsea, MI, 1987.
Kanare, H. M. *Writing the Laboratory Notebook*, American Chemical Society, Washington, D. C., 1985.

11. Method Selection

WHICH ASSAY METHOD SHOULD BE USED?

There used to be just one assay for a given analyte in biological samples. Today there are many. No longer are analysts limited to a few slow biological assays with their inherent poor precision or to chemical methods of limited usefulness due to a lack of sensitivity and selectivity. The analyst today is faced with a dazzling array of sensitive and selective assay methods for the determination of analytes in biological systems. There are methods using electrophoresis, chromatography, mass spectrometry, infrared, visible and ultraviolet spectroscopy, enzymes, antibodies, flow injection analysis, and many others. There is an increasing use of the so-called hyphenated methods, that is, those assay methods that use more than one technique. An example of the hyphenated methods is the combination of the selectivity of antibody binding, the sensitivity of enzymatic amplification, and the quantitation of uv/vis spectroscopy in the ELISA methodologies.

Biochemical/chemical analyses of biological samples are done to answer specific questions. Most of these questions can be answered by determining whether given compounds are present, at what level they are present, and if (and how much) their concentrations change with time. Key questions for an analyst include: What is the problem that needs these analyses? What is the analyte? What is the sample matrix? What are the possibilities that the other components of the matrix will interfere with the analysis? What assay methodology characteristics are required to provide the results needed to solve the problem? Different questions, analytes, analyte concentrations, and matrices require different assay methodologies or combinations of methodologies. The key to a successful solution in the selection of the best assay methodology is an evaluation of the problem at hand, the sample matrices, the available assay methodology, and the resources of the laboratory. Most likely more than one assay method will yield results with the necessary accuracy and precision, and thus method selection should consider secondary factors such as cost and laboratory resources.

THE PROBLEM

Chemical assays are almost never done for their own sake. They almost always are done to aid in the solution of some larger problem. Since the assay method used can have a considerable impact on the specific usefulness of the results and their interpretation for the resolution of these problems, it is important that the analyst understand the nature of the underlying problem. It is usually advisable that an analyst be a part of the initial planning group that defines the protocols for the solution of the overall problem at hand. Under such situations analysts often can suggest modifications in the proposed protocols that will improve the usefulness of the analytical results and decrease costs. Once analysts understand the problem, they can select assays with the appropriate sensitivity, selectivity, limit of detection, and precision.

Typical components of overall problems are shown in Table 11.1. First the analyst needs to understand what will be the end use of the results. If the analyses are used for in-house monitoring, the required documentation; the need for backup samples, agreement between analysts, agreement between laboratories,

TABLE 11.1. General Problem Characteristics

End use of results
Analyte to be measured
Matrix containing the analyte
Analyte concentration range of importance to problem solution
Number of analyses needed
Snapshot in time or trend over time
Required accuracy
Required precision
    By run, by day, by analyst, by laboratory
Data validation requirements
Necessity for acceptance of data
    By peers, legal system, general public, etc.
Needed response time
Need for replicates

method and data validation; and the necessity for acceptance of data by peers, the legal system, and the general public are rather limited. On the other hand, if the analyses are being used for original research, clinical and public health analyses, or litigation purposes, then these requirements and needs can be quite extensive. Often there are significant differences in the needs for levels of accuracy. Identification of the analyte(s) of concern is crucial, and the chemical structure and physical properties of the analyte(s) are the determining factors of what measurement techniques might be used to determine the level of the analyte in a sample.

Different disciplines and different scientists define analytes differently. Some define the analyte by an exact designation of a specific chemical structure. However, others define the analyte by a generic term that good analysts will know refers to a group of different but chemically similar compounds. Analyses of the concentrations of such groups of compounds will usually require measurement of the concentrations of a number of individual compounds. For example, protein concentration is a frequently requested value. There are many different proteins, approximately 5000 species in a single cell. Thus protein assays usually need to be selected for a feature of proteins that is common to all (or most) of the 5000 proteins. Many times an analyte will be defined in terms of its biological activity. Often several chemical compounds will elicit a given biological activity.[1] The analyst thus needs to understand which chemical compound(s) will elicit the biological activities of concern. Then the analytical methods can be selected that will permit the determination of all such compounds. Consideration should be given to the possibility of the existence of multiple components with single biological activities, precursors, and storage forms and their possible relevance to the solution of the problem.

Most problems require a knowledge of the analyte concentration within a limited range, and the analyst needs to know that concentration range. The number of analyses needed can be a crucial factor. Some problems require thousands of analyses and others only a few. The analyst using the assay method must be capable of performing such assays within the time frame and budget of the project. There is a significant difference in the implementation and the needed quality assurance when an analysis is to be done on a one-time basis and when multiple analyses are done to measure trends over time. Thus the analyst will need to know whether the problem requires a snapshot in a single time period or whether trend analysis is a component part of the problem's solution. Some problems require that absolute concentration values be obtained. Others do not need absolute values, but do require that the concentration differences between samples be well characterized. These two situations lead to different requirements

---

[1] For example, the concept of "vitamin" is one of biological activity. There are six primary chemical compounds that have vitamin B_6 activity, and to do a chemical analysis for vitamin B_6 concentration one must measure the concentration of six different compounds.

for method accuracy and precision. The needed precision can vary widely. For example, high precision is needed when measuring calcium in human sera, for the calcium levels are under homoeostatic control and small differences can have significant implications for the assessment of health status. Different problems lead to different requirements for precision between runs, by days, by analysts, and by laboratory. Precision of the results around the limit of detection will have significant effects on the number of reported false positives[2] and false negatives. The need for peer acceptance of the results will vary by the problem and can have significant impact on what are conceived to be acceptable methods and on the data validation requirements.

A key factor in method selection can be the needed response time. Usually the results need to be available within a specified time period. Many assays have relatively long turnaround times, and often the results are needed within a time frame that is shorter that the turnaround time for a given assay methodology. Finally, the need for replicate analyses can have significant impact on the number of analyses as well as the storage capacity of the laboratory for the replicate samples.

SAMPLE MATRICES

Once the analyst understands the nature of the problem requiring the chemical analyses, the characteristics of the sample matrices need to be examined. Typical matrix components of concern are shown in Table 11.2. The analyst will need to determine the maximum known range of analyte concentrations in samples, the range of concentrations found in "routine" samples, and the "natural" variability of samples. Such information will permit the proper choice of the assay methods

TABLE 11.2. **Sample Matrix Characteristics**

Analyte concentrations in samples
    Concentration ranges in "routine" samples
    Maximum known ranges in all samples
    "Natural" variability of concentrations
Probability of sample contamination
Probability of interfering materials in sample matrix
Probability of modifications of analyte with storage and/or processing
Physical state, particle size, and homogeneity of sample
Toxicity of sample
Availability and cost of sample
Use of sample after analysis

---

[2] A false positive is the finding that a sample contains the chemical component when it does not. A false negative is the finding that a sample does not contain the chemical component when it does.

based upon their linear ranges and their usual precision. There are many cases where the sample matrix can be contaminated or the analyte destroyed due to improper storage and processing procedures. The heavy metal contamination of samples by glassware and cutting knives are two well-known examples. The loss of analyte due to the action of naturally occurring enzymes and by oxidation by atmospheric oxygen is common. Storage at low temperatures (less than $-20°C$) will stabilize most samples. Freeze drying is another technique that stabilizes many samples.

As was noted earlier, the nonanalyte components and properties of the matrix are key factors in determining which measurement techniques may not be used to assay the analyte in the matrix. The analyst needs to know enough about the matrix composition to understand which methods cannot be used because the nonanalyte components of the matrix will interfere with the analysis of the analyte. For example, the ninhydrin reagent, which reacts with α-amino groups, should not be used to measure protein in blood sera because blood sera has many nonprotein components that have α-amino groups (see Table 1.4). Many samples are very susceptible to spoilage and must be carefully handled and stored. The analyst needs to have some idea of the stability of the analyte in the sample matrix and the potential for changes in analyte concentration with homogenization and other processing of the sample. While some samples are inherently homo-geneous,[3] most are not. In most circumstances sample homogenization will be necessary so that subsamples have a composition representative of the whole sample. The analyst needs to have information on the physical state of the sample, the particle size, and the inherent sample homogeneity with respect to the analyte. That information along with the needed sample size will determine the characteristics of the required homogenizing process. Homogeneous sample preparation can be quite difficult, and it is astoundingly easy to end up with a heterogeneous sample when using standard compositing processes. Analysts should always seriously consider experimentally demonstrating that the homo-genized subsamples are actually homogeneous at the level of the quantity of sample needed for each determination. Some samples are toxic and can present serious risk to the analyst. For example, the biological hazards of human blood sera due to risk of HIV or AIDs and hepatitis, to mention a few, are well known. Different samples will have different procurement costs, and these costs can have significant impact on the amount of sample that is available for analysis. Finally the analyst needs to know what will be the use of the sample after its analysis. If the sample will be discarded, then destructive sample preparation and analysis systems can be used. On the other hand, if the sample is then to be used in its intended fashion or for further analyses, then nondestructive techniques are needed.

---

[3] Very few samples of biological origin are homogeneous. Most solutions are homogeneous; most solid materials are not; suspensions are frequently heterogeneous.

ASSAY METHODS

Only after the problem is defined, the analytes of concern are identified, and the matrix understood, should the analyst start detailed method selection. Good sources for the identification of appropriate methods include the published chemical literature, vendors of scientific instruments, and laboratories currently doing the analyses. Novices in the field may have a tendency to only turn to the literature to find out what methods are suitable. This is a sound approach and is often suggested in textbooks. However, contacting fellow analysts is often quicker and more efficient than traditional literature searches. It is usually best first to obtain a general knowledge of the available assay methodologies and then to contact fellow analysts, and to obtain their suggestions on the best choices of assay methodologies given the restrictions imposed by the problem, analyte, matrix, and laboratory resources. Talking with one's peers, either by phone or at professional meetings, is usually the contact of choice, since it requires less time of the information provider than lengthy and time-consuming letter writing.

IDEAL METHODS

Method selection is a complex process. A reasonable starting point is to conceptualize what might be the "ideal method." The authors' suggestion of an ideal method was shown in Figure 10.1. There are generally about ten stages to an assay method: sample selection, shipment, and storage; homogenization, subsampling, and storage; separation by chemical class; separation within chemical class; detection; identification; quantification; computation; data validation; and report generation. It is important to note that the accuracy of the final results is dependent upon all the steps of the assay. Chemical analysis of biological samples is a holistic process. Any mistake in any step will lead to an inaccurate final result. The probability of errors canceling each other is quite small.

The potential sources of errors in chemical analyses of biological samples are many and varied. If the wrong samples are selected, then the answers will be wrong. If the sample is shipped or stored improperly, then the answers will be wrong. If the samples are not properly homogenized, then the subsamples are unlikely to be representative of the whole sample. If the separation by chemical classes includes extraction processes that do not completely extract the analyte or if the analyte is partially or completely destroyed during the extraction, then the result will be too low. If the analytes are not separated from interfering compounds during a separation within a chemical class (e.g., chromatography), then the final results may be either low or high, depending upon the characteristics of the detection system. If the detection system is not selective, lacks sensitivity, is nonlinear, drifts, or measures compounds other than those desired, then the results will be incorrect or imprecise, or both. The computation of the final results all too frequently is a major source of error. Computation errors tend to be quite large, frequently orders of magnitude. Errors made in the data validation steps may lead to errors in the computations and certainly have the potential for causing an analyst to accept results that should be rejected or vice versa. Even the development and preparation of the final reports from the results can have errors. Errors made at these stages are just as damaging as those made earlier in the analysis process.

The potential for error in individual analyses and the frequency of such errors is so common that the authors believe that all analyte assay data sets must be individually validated. Detailed discussion of data validation is presented in Chapter 10. However, it should be noted that aspects of the quality-control procedures for data validation should be built into the assay method protocol as it is developed. Quality-control procedures can rarely be successfully tacked onto a method protocol after it has been completely developed.

EVALUATION OF METHODS

Typical method characteristics of concern are shown in Table 11.3. Unfortunately, many scientists and analysts assume that because an assay was satisfactory for other analysts in analyzing the analyte in other matrices, the method will be satisfactory for all occasions. This is a dangerous assumption. Even when well-documented assays are selected, matrix effects can be significant and lead to serious errors for an unwary analyst. For example, cholesterol in blood serum can be accurately determined using the enzyme cholesterol oxidase, and this is a standard assay for blood sera. However, when this assay is used to determine cholesterol in human diets, the results are about 20% too high because the enzyme cholesterol oxidase is actually a sterol oxidase, and the common plant sterols as well as cholesterol give a positive response in this assay. Plant sterols are not found in blood serum, and thus they do not interfere in that assay, but they do interfere when human diets are analyzed.

Thus, once an assay methodology has been tentatively selected, it should be specifically validated (Table 11.4) for the analyses of desired analytes in the current sample matrices. Such validations will alert the analyst to possible problems due to matrix effects, and successful validations will give credibility to the analytical results. The analyst should first analyze a series of standards of known concentration over the range of the analyte concentrations that is dictated by the problem for which the assay is needed.[4] The analyses of these standards

---

[4] If the accuracy and precision are acceptable within this range, it matters little that the methodology results are inaccurate or imprecise at concentrations outside that range.

TABLE 11.3.   Method Characteristics

Accuracy
   Selectivity
Precision
Figures of merit
   Correlation coefficient (R^2)
   Sensitivity
   Linear range
   Limit of detection/quantification
Level of method validation
   In house, published in peer-reviewed literature, published in
        methods book, published by trade, professional, or
        standards organization, statutory publications
Built in data validation
Fragility/ruggedness
Cost factors
Peer acceptance
Safety features
Time factors

TABLE 11.4.   Method Validation

Analysis of standards
Method of standard additions
Analysis of SRMs
Comparison with accepted methods
Analysis of pool samples
Review

should yield acceptable figures of merit and precision of results. The use of pure standards alone to check the accuracy and precision of an assay method is not usually sufficient. Thus, once the standard analyses yield acceptable results, the analyst should analyze the same set of standards in the presence of the matrix (method of standard additions, spiking). The entire assay protocol should be used in this evaluation. The selected assay method should also give a linear response to increasing concentrations of analyte with acceptable accuracy, precision, and figures of merit when the methods of standard additions is used. Care should be taken that the spikes have the same chemical structure as the analyte found in the sample. If more than one analyte is expected in the sample matrix, there should be a series of spikes for each analyte believed to be in the sample. The goal is to demonstrate that nonanalyte compounds known to occur in the matrix, and that have the potential to interfere with the assay, do not interfere with measurement of analyte. The precision should be checked with at least several sets of pool samples containing low, medium, and high levels of the analyte.

Once acceptable results are obtained for the method of standard additions, then the analyst should analyze an appropriate quality-control material (QCM)[5] to demonstrate that the method under consideration can yield the known concentration(s) of the analyte in standard reference material within the specified tolerance limits of the QCM. Appropriate quality-control materials are those with matrix and analyte concentrations similar to samples to be assayed. The quality-control materials, calibration standards, and internal standards should have been previously prepared in batches according to SOP and their content documented by analyses. Often Standard Reference Materials are not available for the desired analyte and matrix combination. In such cases, the analyst will need to demonstrate that the method under consideration will give identical results to an accuracy-based method (see below). This demonstration that the method under consideration gives the "correct" answer for a standard reference material or that it gives "identical" results when compared with an accepted assay methodology is crucial for peer acceptance of the analytical data. Failure to demonstrate such compatibility will most likely result in peer rejection of the analytical data. Finally, the analyst needs to analyze a series of pool samples (low, medium, and high concentrations of analyte) to determine the precision of the assay within and between assay batches on a single day and day by day. These analyses will establish the expected precision of the assay. Such analytical data for pool samples can also be used to establish the database for the quality-control charts that will be used in the data set validations.

When chromatographic or other separation techniques are used in the assay methodologies, internal standards should be developed and included as part of the methodology. Given the potential for misidentification in complex mixtures and difficulty of reproducible sample introduction in chromatographic systems, most analysts will not accept results from chromatographic analyses that do not contain internal standards. Internal standards should be of similar chemistry to the analytes of concern and should be added in concentrations similar to those of the analyte in the sample matrix. It is crucial that the internal standard not coelute with components in the sample matrix, and this aspect of the behavior of the internal standard must be experimentally evaluated. It is best to add the internal standard as early as possible in the assay procedure. If it is not possible to use internal standards, then external standards may be used, but they are not nearly as acceptable.

Since result computations are frequent sources of analytical error, the analyst validating an analytical method should use certified algorithms and computer

---

[5] Ideally the quality-control material will be a Standard Reference Material produced by the National Institute of Standards and Technology (NIST) (best) or a commercial reference material tied to a Standard Reference Material (next best). Often the analyst has to use a pool sample that has been either analyzed by another laboratory (all right) or has been analyzed in house by an accuracy-based method (barely acceptable).

programs whenever possible and should always cross-check that computation algorithms yield the correct results. Once it has been demonstrated that the computation process is error free, the program should be frozen and an archival set saved. The analyst doing the routine analyses should not be able to modify these computation programs. Many laboratories participate in the analyses of "round robin" check samples as a cross-check of the their analytical abilities. In "round robins" homogeneous check samples of known concentration (but not known to the analyst) are sent to a series of laboratories for analyses. The results are shared across all the laboratories. In this fashion each laboratory can ascertain how well it does in relation to the other laboratories. Care must be taken not to let the analyst know which individual sample is a check sample, because there is a tendency to give special attention to a check sample. Some analysts even make a mini-research project of the analyses of check samples. This is not the way to check the routine performance of a laboratory or an analyst.

METHOD MODIFICATION

Since many assay methods were developed for selected sample matrices, analysts frequently will need to do some method development or modification to fit the method to their unique sample matrix. It is quite common that even after a method has been chosen, it will be necessary for the analyst to make some further modifications in the procedure. This is due to differences in the problem, samples, or resources associated with the original technique and those associated with the analyst's current situation. Sometimes the analyst will conclude that no existing method will suffice and a new method will be needed. The entry point into the development of a new or modified method depends upon the state of the field, the nature of the problem, and the expertise and wishes of the analyst. Figure 11.1 is a suggested flow chart for the development of an ideal analytical method. As can be seen, there are many entry points. This flow chart also demonstrates the interdependence of the different parts of the method and its development. Note that all aspects of the quality-control procedure are built into the method development. This is as it should be for optimal method development.

FRAGILITY/RUGGEDNESS OF A METHOD

Not all methods show the same variations on a day-to-day basis, an analyst-to-analyst basis, or a laboratory-to-laboratory basis. Some methods work almost all the time, and others only when extraordinary care is taken. Methods that yield the same accuracy and precision no matter who the analyst is, or what day it is, or in what laboratory the work is done can be said to be "rugged." Those that do not

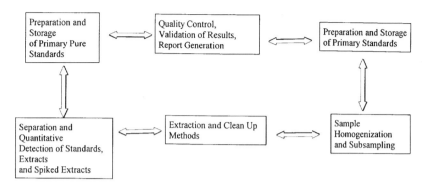

Fig. 11.1.   Idealized method development.

can be called "fragile." Obviously, rugged methods are preferred if all other things are equal. Information on the degree of ruggedness may be difficult to acquire from the literature or manufacturer but is often readily available from other analysts who are using the method. The extra time spent in learning about the ruggedness of a method can save a significant amount of time over the term of the project.

COSTS

In today's budget-minded atmosphere, the cost of assay methodologies and the cost differential between different assay methodologies can be key factors in the selection of a method. The critical factor in the evaluation of the cost is whether or not the results of the analyses will be worth the effort and resources expended on them.

The costs of an assay depend upon the cost of labor, instrumentation acquisition and maintenance, reagents, computation facilities, and laboratory physical facilities. Labor costs are usually the single greatest component of the cost of analyses. Some assay methods require a very high level of technical training, and others require almost no training. Assays that require a Ph.D. as an analyst will probably be more expensive than those that require an analyst with less than four years of high school. Further specialized training is frequently required, and the cost and time required for this training should be considered. If the analyses will be done over a long period of time, perhaps more than one analyst should be trained so that shutdowns can be limited when the primary analyst leaves or becomes incapacitated. There are striking differences between the purchase costs of different types of instrumentation as well as their maintenance costs. Consideration should be given as to whether or not new equipment will need to be purchased or leased or whether the existing equipment can be used. Consideration of the costs required for equipment maintenance is often overlooked. Repair personnel often have to travel to the laboratory site to repair the instrumentation. Repair cost estimations should include the minimum and average cost per visit and the normal delay period before the repair person appears. Some reagents are literally worth more than their weight in gold, and others are quite inexpensive. The cost of chemical disposal can be appreciable with some types of assays. Physical facilities costs include those for humidity and temperature control, for special air handling facilities, for an adequate electrical and lighting systems, and for antivibration benches. Clean room facilities are sometime required. Some assays will require the installation of safety features. Most modern instruments have computer components, and the analyst should be sure these computer systems can be reliably used in the laboratory. Typical problems for computer operations are the result of dirt, vibration, moisture, the lack of proper grounds, and electrical noise. An astounding number of pieces of equipment generate electrical spikes that incapacitate computers. The costs of the computational and report generation systems are often overlooked, but many modern assay systems require high-speed computer facilities with rather large memory capacities.

SAFETY

As society in general and analysts specifically become more aware of hazards in the workplace and of the fragility of our environment, the safety components of analytical procedures are receiving more attention. All components of any assay should be carefully evaluated for the potential danger to the analyst, to other members of a laboratory, and to the environment and the general public to ensure that necessary safety features have been incorporated. The dangers of radioactive material are well known, but other hazards are not as well recognized. Many reagents, some solvents, and some biological matrices are known to be dangerous because of their carcinogenic, mutagenic, toxic, explosive, and flammable attributes. The days when it was acceptable to flush toxic materials down the drain are well behind us. Since most of these materials are as toxic in the general environment as they are in the laboratory, specialized disposal systems must be used. Unfortunately, disposal of chemical wastes has become very expensive. Some instruments present electrical hazards (e.g., high-voltage electrophoresis), and others have hazards such as the potential for radioactive exposure and exposure to laser beams.

TIME FACTORS

There are three key issues in evaluating the time factors of an assay: the turnaround time for an individual analysis (i.e., how long will it take to get the results of an individual analysis?), the throughput time (i.e., how many samples can be analyzed in a given time period?), and implementation time (how long will it take to have the assay method up and running in the laboratory?) Acceptable turnaround times are usually a function of the problem for which the analyses are being done. If a product has a lifetime of two days or if a critical medical decision needs to made immediately, then analyses that take a week are obviously not acceptable. On the other hand, there is no need for really rapid analyses if the data will not be used within a month. When the turnaround time is computed, calculate the total time required for the analyses. All too often analysts only include the measurement step in an assessment of the time required and forget to include other steps such as sampling, homogenization, and computation, which actually may take more time than the actual measurement of the signal. For problems that require large numbers of analyses, throughput may be even more important than the turnaround time for an individual analysis. Implementation time should include time for purchase and receipt of the instrumentation, the reagents, any needed modification of the facilities, and any required analyst training.

QUALITY CONTROL

Today's assay methodologies are sufficiently complex that anyone can make a mistake in performing an analysis. The selected assay methodology should have a documented, built-in data set validation quality-assurance system utilizing pure standards, reference material, and internal standards or the equivalent. For long-term studies it will be necessary to demonstrate that the methodology does not drift over the period of study. A detailed discussion of data validation and assay quality control is given in Chapter 10.

PEER ACCEPTANCE

In some analytical situations appearances are as important as the actual results. It is often quite important that the analytical result be believed. This is frequently the case for those analyses that are performed in response to health and legal issues. One of the key requirements for producing believable data is the peer acceptance of the method. Acceptance of a method by the analytical chemistry community usually requires that the general assay methodology be published in a peer-reviewed journal, that the method has been given significant exposure to the peer audience, and that the published restraints for the general assay are met in the protocols for the analysis of the specific sample matrix. Demonstrating that the methodology has gained necessary peer acceptance may even require that the method has been found acceptable after having undergone evaluation in round robin studies.

ACCURACY-BASED MEASUREMENT SYSTEMS

Typically the assay method chosen for an assay is selected because it can be performed rapidly and at low cost. Such methods have been called "field methods." Unfortunately such rapid, low-cost "field methods" can yield inaccurate results if insufficient care is taken in their selection, and it is not uncommon for different field methods to give different assay results for the same sample. However, definitive chemical assays (such as isotope dilution mass spectrometry) are not appropriate for routine analyses since they are very time consuming, are expensive, and require exceptional technical skill. The conflict between analysis cost and assay accuracy is a real and recurring issue in the selection of the appropriate assay method. The approach suggested by Tiez[6] is a model worthy of consideration in the validation of field methods for the analysis of given analytes in given matrices. The concept is that there should be "traceability" between the selected field methods and the definitive methods, as shown in Figure 11.2. A "definitive method" is one that is capable of providing the highest accuracy of all methods for the determination of a specific analyte in the matrix; a "reference method" is one that completely documents accuracy and precision and has a low susceptibility to interfering compounds that are found in the sample matrix; a "field method" is one used for routine determinations of the analyte in the matrix. Each of these methods has a corresponding refer-ence/control material. A "primary reference material" is a well-characterized, stable, homogeneous material in which the concentration of the analyte has been established by a definitive method or two independent[7] assay methodologies of known high accuracy. Secondary reference materials are lower-cost materials that are selected or prepared to simulate the matrices and analyte concentrations found in the primary reference materials. Control materials[8] are usually made from pooled samples obtained locally and are used in the quality control of the routine analysis of samples. The systematic use of these methods and materials is called the accuracy-based measurement system. Demonstration that the method is an accuracy-based method is becoming a common requirement for peer acceptance of assay methodologies.

---

[6] R. W. Tiez, *Clin. Chem* **25**, 833– 839 (1979).

[7] *Independent* in this case means that the two methods are founded upon measurements based upon different chemical principles.

[8] See Table 11.5.

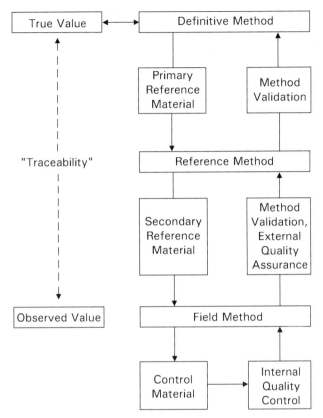

Fig. 11.2. Accuracy based measurements. Taken from "Text Book of Clinical Chemistry" 2nd Edition, edited by C. A. Burtis and E. R. Ashwood. Publisher W. B. Saunders Co. Reprinted with permission.

TABLE 11.5.   Uses of the Components of Accuracy-Based Systems

Methods	
Definitive method	Used in evaluation of reference method and in evaluation of primary reference material
Reference method	Used to assign values to secondary reference materials and control materials; used in the evaluation of field method
Field method	Used for routine analysis of samples and control material
Materials	
Primary reference material	Used in development of reference method and in calibration of secondary reference material
Secondary reference material	Used as working standards for field methods and used to assign values control materials
Control material	Used for quality control of routine analysis of sample

Taken from "Text Book of Clinical Chemistry" 2nd Edition, edited by C. A. Burtis and E. R. Ashwood. Publisher W. B. Saunders Co. Reprinted with permission.

AVAILABLE RESOURCES

Once the problem, matrix, and method characteristics have been defined, the analyst should evaluate what resources are available in the laboratory for the analyses. Those resources include funding, equipment and computer facilities, personnel capabilities, and space. Generally speaking, it is less expensive and faster to select methodologies that use the existing resources of the laboratory rather than to select a method that requires acquisition of novel equipment and retraining of personnel or hiring of new personnel. It is usually less expensive and less time consuming to purchase analytical instrumentation than to make such instrumentation.

METHODS OF CHOICE

Table 11.6 gives the authors' selections for the separation and detection methods of choice for the analysis of some of the more common analyte classes found in biological samples. Obviously, others may use different techniques for the analysis of specific compounds in specific matrices; however, we feel that these techniques deserve serious consideration when the analyst is selecting a method for some analysis program. Note that there are still some serious inadequacies in the available analytical procedures. For example, at the present time (2000) there

TABLE 11.6. Authors' Choices of Assays by Chemical Class

Compounds	Separation of Choice	Detection of Choice
H_2O	GLC	Thermal conductivity
Inorganic ions	Ion chromatography, capillary electrophoresis	Conductance
Metals	Digest	Atomic absorption, emission spectroscopy
Amino acids, etc.	HPLC	uv/vis spectrophotometry, electrochemical, fluorescence, chemiluminescence
Fatty acids, etc.	GLC	Flame ionization (FI), mass spectrometry
Carbohydrates	HPLC, GLC	FI, refractive index (RI), enzyme assays
Vitamins, coenzymes, isoprenoids, porphyrins	HPLC	uv/vis spectrophotometry, electrochemical, fluorescence, chemiluminescence, mass spectrometry
Lipids and phospholipids	TLC	Chemical detection
Proteins	HPLC, electrophoresis	Chemical detection reactions, enzyme assays, immune assays
DNA and RNA	Electrophoresis	Chemical detection reactions, uv/vis spectrophotometry, fluorescence
Polysaccharides	?	Chemical detection reactions, immune assays

appear to be no generally acceptable methods for the generic separation of polysaccharides. Furthermore, the analysis of the concentrations of transition metals in biological samples almost always requires the digestion of the sample to its constituent elemental components; information of the speciation of the metal components of the sample is thus invariably lost.

Appendix 1   Units

All laboratory measurements have units. Given that there are many different unit systems, it is obviously possible for significant confusion and miscommunication to occur unless a discipline uses a standard unit system. The chemical community uses both the International System of Units (SI), Table A1.1, and the metric system of units.

TABLE A1.1.   The Fundamental Units of Measurement (SI)

Property	Name of Unit	Symbol
Length	meter	m
Mass	kilogram	kg
Time	second	s
Electric current	ampere	A
Temperature	Kelvin	K
Luminous intensity	candela	cd
Amount of substance	mole	mol

METRIC CONVERSIONS FROM SI UNITS

$$1 \text{ angstrom } (\overset{\circ}{A}) = 1.0 \times 10^{-8} \text{ cm} = 10 \text{ nm}$$

$$1 \text{ liter (L)} = 1000 \text{ cubic centimeters } (cm^3) = 1000 \text{ milliliters (mL)}$$

$$0^\circ \text{ K} = -273.15^\circ C$$

Many times the quantities used or measured in an experiment are several-orders-of-magnitude greater or less than the standard SI or metric units. Such numbers can be awkward to use, and numerous errors occur in their use. Two conventions have been developed to simplify the communication of such large or small units. One is the use of *scientific notation* and the other is the use of *standard prefixes*.

Scientific Notation. With scientific notation values are represented in two parts. The first part contains the digits, which are expressed as significant digits,[1] and the second are the powers of ten of the value.

Example 1. You have 3200 mL of solution, and your measurements are good to two significant digits. In scientific notation, this value would be presented as 3.2×10^3 mL.

Example 2. You have 0.00450 mL of solution, and your measurements are good to two significant digits. In scientific notation, this value would be presented as 4.5×10^{-3} mL.

Standard Prefixes. The other approach to the presentation of quantities that are several-orders-of-magnitude greater or less than the standard SI or metric units is to change the reporting units. Some of the standard prefixes are shown in Table A1.2.

Amounts

Chemical analysts express amounts in either grams (g) or moles (mol). The number of moles in a sample equals its weight in grams divided by its molecular weight. There are 6.023×10^{23} molecules per mole. The number of grams in a given volume of solution equals its concentration (g/L) times the volume of the sample (L); likewise the number of moles in a given volume of solution equals its

---

[1] The significant digit convention says that an analyst reports all digits known with certainty and the first one that is uncertain. The significant digit convention is discussed in detail in Appendix 3.

TABLE A1.2. Some Standard Prefixes

Prefix	Abbreviation	Meaning	Example
kilo	k	10^3	1000 m = 1 km
			1000 g = 1 kg
deci	d	10^{-1}	0.1 L = 1 dL
centi	c	10^{-2}	100 cm = 1 m
milli	m	10^{-3}	1000 mL = 1 L
			1000 mg = 1 g
micro	μ	10^{-6}	1000 μL = 1 mL
			1000 μg = 1 mg
nano	n	10^{-9}	1000 nL = 1 μL
			1000 ng = 1 μg
pico	p	10^{-12}	1000 pmol = 1 nmol

concentration [M (mol/L)] times the volume of the sample (L). Thus the amount in solution is a function of both concentration and volume.

$$amount = concentration \times volume$$

$$g = (g/L) \times (L)$$

$$mol = (mol/L) \times (L)$$

$$1 \text{ millimole (mmol)} = 1 \times 10^{-3} \text{ mol}$$

$$1 \text{ micromole (μmol)} = 1 \times 10^{-3} \text{ mmol}$$

$$1 \text{ nanomole (nmol)} = 1 \times 10^{-3} \text{ μmol}$$

$$1 \text{ picomole (pmol)} = 1 \times 10^{-3} \text{ nmol}$$

$$1 \text{mole} = 1000 \text{ mmol}$$

$$1 \text{ millimole} = 1000 \text{ μmol}$$

$$1 \text{ μmicromole} = 1000 \text{ nmol}$$

$$1 \text{ nanomole} = 1000 \text{ pmol}$$

$$1 \text{ g} = 1000 \text{ mg}$$

$$1 \text{ mg} = 1000 \text{ μg}$$

$$1 \text{ μg} = 1000 \text{ ng}$$

$$1 \text{ ng} = 1000 \text{ pg}$$

Concentrations

The base units for concentration are either molar, abbreviated M or grams per liter (g/L). A 1 M solution has a concentration of 1 mole per liter. Note that concentration is not dependent upon volume or amount. The common unit conversions of concentrations are:

$$1 \text{ M} = 1 \text{ mol/L} = 1 \text{ mmol/mL} = 1 \text{ μmol/μL}$$

$$1 \text{ mM} = 1 \text{ mmol/L} = 1 \text{ μmol/mL} = 1 \text{ nmol/μL}$$

$$1 \text{ μmolar (μM)} = 1 \text{ μmol/L} = 1 \text{ nmol/mL} = 1 \text{ pmol/μL}$$

$$1 \text{ g/L} = 1 \text{ mg/mL} = 1 \text{ μg/μL}$$

$$1 \text{ mg/L} = 1 \text{ μg/mL} = 1 \text{ ng/μL}$$

$$1 \text{ μg} = 1 \text{ ng/mL} = 1 \text{ pg/μL}$$

Always put units on data. Not to do so will result answers being meaningless! It is a serious mistake to use incorrect or no units.

Unit Analysis

To check whether or not analysts are doing the computations correctly and have the correct units, they should use a technique called unit (or dimensional) analysis. For example, in spectrophotometry Beer's law is: $A = \varepsilon Cl$. The units of each component are as follows:

A = dimensionless or sometimes absorbance units (AU)

C = concentration (M)

l = (cm)

$\varepsilon = (\text{M}^{-1} \text{ cm}^{-1})$

Unit analysis would consist of the following. The formula for Beers law is $A = \varepsilon Cl$. Thus the units for $A = \text{M}^{-1} \text{ cm}^{-1} \text{ M cm}$ = no units or AU. Or for the computation of ε would be

$$\varepsilon = A/(Cl) = \text{AU}/(\text{M cm}) = \text{AU M}^{-1} \text{ cm}^{-1}$$

However, experimentally if the concentration were given in g/L, then it would be obvious that some computation was needed to obtain the correct units. For example, ε was to be reported in AU M^{-1} cm^{-1}, but in the unit analysis one would get $\varepsilon = AU/[(g/L)cm]$. This would indicate that another computation step would be needed to get the correct answer. Since $M = (g/L)/MW$, and MW has units of g/mol, the step would be as follows:

$$\varepsilon = AU/[(g/L)cm/MW] = AU\ M^{-1}\ cm^{-1}$$

Appendix 2   Statistics for Chemical Analyses in Biological Systems

As far as the laws of mathematics refer to reality, they are not certain; as far as they are certain, they do not refer to reality.

Albert Einstein

Chemical measurements in the real world always have the potential of containing errors.[1] The practical experience of the authors is that all chemical analyses of biological samples have errors. It has been stated that " ... No quantitative results are of any value unless they are accompanied by some estimate of the errors inherent in them."[2] The authors agree and suggest that honest presentations of the results of any chemical analysis of biological systems will contain an estimate of these errors. The purpose of this section is to provide a rudimentary set of simple statistical tools for the estimation of these errors. The reader is referred to standard statistics textbooks for the mathematical and/or statistical explanations

---

[1] Errors are not necessarily the same as mistakes. The word *error* in this discussion includes gross mistakes and the statistical concepts of bias and random errors.

[2] J. C. Miller and J. N. Miller, *Statistics for Analytical Chemistry,* 2nd edition, John Wiley and Sons, NY, 1988.

of these procedures as well as for a more sophisticated discussion of these issues. The goal is to provide objective ways of assessing analytical data so that the following key questions can be addressed.

Key Questions:

What is the best estimate of true value of the concentration?
What is the confidence in the estimate of the true value?
Do the two samples have the same concentration?
When can an outlier be rejected with some confidence?

ERRORS

There are three types of errors; *gross errors* resulting from serious malfunctions of one or more components of the assay system, *bias* (systematic errors that affect the accuracy of results), and *random errors* (resulting from random fluctuations in the concentration of the analyte in the sample and/or in various components of the assay process that affect the precision of results). Gross errors are usually easily identified by the analyst in observations of system failure (e.g., obvious contamination or instrument failure), and such identification should lead to the rejection of the entire analysis set and a repeat of the analysis of that set of samples. The identification of outliers (see below) is often used to identify less obvious gross errors. The bias (inaccuracy) of a method can be difficult to determine. However, there are statistical tests for the estimation of the possible biases between two different assay methodologies, and the reader is referred to a statistics text such as the one listed at the end of this section. All measurements have a degree of randomness, and the estimation of this randomness is crucial in the determination of the confidence that should be given the analytical value of the quantity being determined.

What Is the Most Probable True Value for the Composition of this Sample?

In the estimation of the most probable value, the underlying assumption is that random error will result in a series of measurements normally distributed around the mean. Figure A2.1 is a graphical representation of an infinite number of normally distributed results. When the results are normally distributed, the most

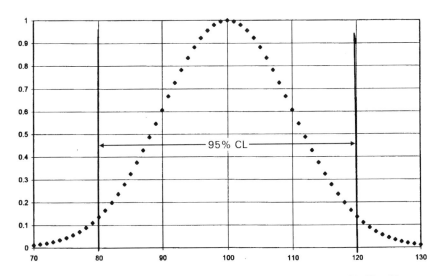

Fig. A2.1. 95% Confidence limits for a Normal Distribution, Mean = 100, SD = 10.

probable value for them is the mean of the results of the assay. The equation for the computation of the *arithmetic mean* ($\bar{x}$) of a data set is:

$$\bar{x} = \frac{\left(\sum_i x_i\right)}{n} \tag{A2.1}$$

where x_i is defined as a value for each individual measurement and n as the total number of these measurements.

The *standard deviation of the mean* of the measurements defines the spread of a set of analytical results. The standard deviation of the mean (SD) is defined:

$$SD_{mean} = \left(\sum_i (\bar{x} - x_i)^2 / (n - 1)\right)^{1/2} \tag{A2.2}$$

Potentially, each step of an analytical assay has variation, and thus the total variation of a measurement is a function of the variation in each step. It is current dogma that the squares of the standard deviations of the means are additive:

$$SD_{total}^2 = SD_{step\ 1}^2 + SD_{step\ 2}^2 + SD_{step\ 3}^2 + \cdots + SD_{step\ N}^2 \tag{A2.3}$$

Note that the magnitude of the standard deviation is dependent upon the units of the results. However, it is often useful to compare different data sets and/or methodologies without regard to the units of the analysis systems. The *relative standard deviation* (RSD)[3] can be used to make such comparisons, since it is a unitless presentation of the standard deviation. The RSD is defined as:

$$\text{RSD} = 100(S.D./\bar{x})(\%) \tag{A2.4}$$

Note that it is expressed as a percent.

The range of values within which the "true value" will fall 95% of the time is called the 95% confidence limit. While the exact calculation of the confidence limit is complex and is better discussed in a statistics course, a crude estimate of the 95% confidence limit is given by:

$$95\% \text{ confidence limit} = \bar{x} \pm 2 \text{ SD} \tag{A2.5}$$

A 95% confidence limit for a data set is shown in Figure A2.1. When an analyst assays several aliquots of a sample and obtains a mean value for the concentration of an analyte, the actual value of that sample's concentration is considered to be between (mean−2 SD) and (mean + 2 SD) with a probability of 0.95. Said another way, the 95% confidence limits for a mean value are from (mean− 2 SD) to (mean + 2SD).

Comparison between Two Experimentally Determined Means

A frequent question asked by the analysts is "Do these two samples have the same concentration?" The question can be answered by comparing the 95% confidence limits for the results of the analysis of each sample.[4]

As a first approximation, if the 95% confidence limits of the analytical results overlap each other, then the values cannot be said to be different. If the 95% confidence limits of the results do not overlap each other, then the values can be said to be different. Figure A2.2 shows the distributions for two sets of results where the analyst can say the samples are different. Figure A2.3 shows the distributions for two sets of results with the same means as in Figure A2.2 but with larger standard deviations. In the latter case, the analyst cannot say the samples are different. The reader is referred to statistics texts for more sophisticated comparisons of analytical results. Pragmatically, however, students can still make a quick estimation of whether or not two means are different.

---

[3] Sometimes called the coefficient of variation (CV).

[4] This discussion of the statistics of measurements has been simplified. The reader is advised to consult a good statistics textbook for a more detailed discussion of the statistics of chemical measurements.

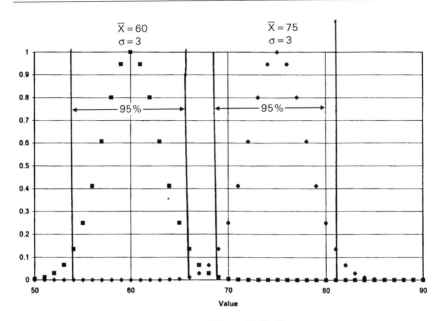

Fig. A2.2. Different distributions.

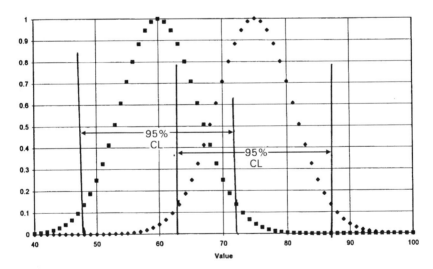

Fig. A2.3. Can not say distributions differ.

Are They Different?

If a mean were reported to be 6540 ± 25, the reader would know that 95% of the time the actual value of the mean will be between the mean-2 SD and the mean + 2 SD. When comparing two means to determine if they were different, compare the 95% confidence limits of the two means.

Example 1.   Are the following two means different?

$$6540 \pm 25 \quad \text{and} \quad 6510 \pm 15$$

The 95% range for the first mean is from

$$6540 + 50 \text{ to } 6540 - 50, \quad \text{i.e.,} 6590 \text{ to } 6490$$

The 95% range for the second mean is from

$$6510 + 30 \text{ to } 6510 - 30, \quad \text{i.e.,} 6540 \text{ to } 6480$$

The 95% confidence limits overlap, thus these two means cannot be said to be different.

Example 2.   Are the following two means different?

$$6500 \pm 25 \quad \text{and} \quad 6590 \pm 15$$

The 95% range for the first mean is from

$$6500 + 50 \text{ to } 6500 - 50, \quad \text{i.e.,} 6550 \text{ to } 6450$$

The 95% range for the second mean is from

$$6590 + 30 \text{ to } 6590 - 30, \quad \text{i.e.,} 6620 \text{ to } 6560$$

The 95% confidence limits *do not* overlap; thus these two means can be said to be different.

Significant Figures

It is obviously useful if the reported analytical results clearly indicate whether or not different results are the same or are different. Proper use of significant figures provides the needed clarity of reporting so that the reported results clearly reflect the degree of confidence that can be given them. Remember that all analytical results should be reported as the *mean ± the standard deviation*. When using significant figures, the basic rule is that *means should contain all the digits known with certainty and contain only the first uncertain digit.* The standard deviations should be reported with one more digit than the means.

When using significant figures, zeros to the right of the decimal point and to the right of the digits are considered part of the significant figures (e.g., 0.120 has three significant figures; the zero to the left of the decimal point is a place holder). Zeros to the right of the decimal point but to the left of the digits are not considered part of the significant figures (e.g., 0.0012 has two significant figures). Zeros to the left of the decimal point and to the right of the digits are usually, but not always, considered part of the significant figures (e.g., 120) has two or three significant figures. These rules are summarized in Table A2.1.

Selecting the Significant Figures of a Result

There is approximately a 95% probability that an individual value will be within the range of (mean − 2 SD) to (mean + 2 SD). Thus the uncertainty of the mean is approximately two times the standard deviation. We can use these statistics to select the significant figures of a result.

Using these concepts, a rule of a thumb for determining the number of decimal places of the mean of the assay is: Multiply the standard deviation by two, then note the decimal place of the first nonzero digit of the product; report the mean to that decimal place and provide one more decimal place for the standard deviation.

$$\text{Certain} \quad \text{Uncertain} \qquad \text{Certain} \quad \text{Uncertain}$$

$$111.22 \pm 0.222 \qquad 111.22 \pm 11.11$$
$$\text{SD}^m \times 2 = 0.44 \qquad \text{SD}^m \times 2 = 22.22$$

Report 111.2 ± 0.22 **Report** 110 ± 11

This rule is easy to use. If the assay yielded a value of 111.22 ± 0.222, then we multiply 0.22 by 2 and obtain 0.44, and thus the results would be reported as 111.2 ± 0.22. Thus the mean is accurate to one decimal place and the standard deviation should be reported as accurate to two decimal places. Alternatively, if the same mean had a standard deviation of 11.11, multiplying the standard

TABLE A2.1. Some Significant Figures

Mean	Significant Figures
12	2
1.2	2
0.120	3
12.0	3
0.0012	2
120	2 or 3

deviation by 2 would yield 22.22, the first nonzero digit would be in the tens place, and the mean would be reported as 110 ± 11.

Use of RSD to Determine Significant Figures

The relative standard deviation (see Eq. A2.4) can also be used to determine the significant figures of a data set. A 1% RSD implies that the mean has a relative uncertainty of 1 part in 100 and thus the number of digits known with certainty would be two and there would be three significant figures. Thus data with a 1% RSD would be reported as 440 ± 4, 1110 ± 12, 0.0210 ± 0.0002, etc. The rules shown in Table A2.2 can be used to estimate the significant figures from the RSDs.

Knowledge of Assay Characteristics and the Selection of Significant Figures

If the relative standard deviations are not known, the data should be reported with significant figures appropriate for assays used to obtain the value. Most chemical assay methods for biological systems have precisions between 2 and 10% RSD. Thus the results of most chemical analyses of biological systems should have only two or three significant figures.

Standard Deviations of Results Computed from Standard Curve Regression Analyses

The standard deviation of the value computed for an unknown usage of a result from the regression analysis of a standard curve includes both the variance of the results of the unknown and the variance of the results of the standard curve. Details of the computations of these standard deviations are presented in Appendix 3.

Outliers

It is not uncommon for data sets to have one or more values that seem unreasonable. Such unreasonable results are called *outliers* and are usually caused by analysts' mistakes and/or instrument malfunction. The presence of outliers in the data sets can result in erroneous means and inflated standard deviations; so it can be desirable to eliminate them from data sets prior to the computation of the

TABLE A2.2. RSD and Significant Figures

RSD	Variance 1 part in	Significant Digits
>0.1%	1000	4
>1.0%	100	3
>10%	10	2
>20%	5	1

TABLE A2.3. Critical Values for the Q Test for the Rejection of Outliers

Number of Samples	Critical Value
4	0.831
5	0.717
6	0.621
7	0.570
8	0.524
9	0.492
10	0.464

Source: E. P. King, *J. Am. Statist. Assoc.* **48,** 531, 1958.

standard statistics. The questions are: "When should a value be considered an outlier and thus rejected?" and "When should a value be considered a normal random variation of the data and be accepted?" *Dixon's Q test* is used to assess the suspect value:

$$Q = \text{absolute value of} \frac{\text{suspect value} - \text{nearest value}}{\text{largest value} - \text{smallest value}} \tag{A2.6}$$

The Q value for a randomly distributed set of results for different sample sizes is shown in Table A2.3. If the computed Q value is larger than the critical value in the table for that sample size, then the test states that the suspect value can be rejected; if not, then the suspect values should be retained.

GENERAL REFERENCE

J. C. Miller and J. N. Miller, *Statistics for Analytical Chemistry*, 2nd Ed., Ellis Harwood, Chichester 1988.

Appendix 3   Standard Curves and Linear Regression Analyses

Quantitative chemical measurements in biological systems almost always require calibration of the relationship of the signal being measured to the concentration or amount of the analyte being determined. Thus analysts typically prepare a sequence of known concentrations of the analyte (called standards) and then measure the signals corresponding to each concentration of the analyte. The signals (dependent variables) are then plotted on the y axis versus the known concentrations (independent variable) on the x axis (see Fig. A3.1).

Definition: A response curve is a plot of the assay response signal versus the analyte concentration or amount.[1]

Once a response curve is established for a given analysis, the signals for samples with unknown concentrations are measured. The unknown concentrations can be determined graphically by drawing a straight line from the observed signal parallel to the x axis over to the response curve and then dropping a straight

---

[1] A response curve is sometimes called a calibration curve.

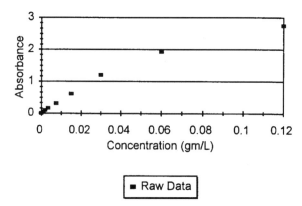

Fig. A3.1. A response curve. (Note the convention is to plot actual data as points and computed data as lines.)

line parallel to the *y* axis down to the *x* axis and reading the concentration (see Fig. A3.2).

However, such graphical computations lack precision and can be tedious if one has multiple samples of unknown concentrations. More sophisticated analysts fit the data from the response curve to some type of mathematical equation. The simplest of such mathematical equations is the equation for the straight line:

$$y = mx + b \qquad\qquad (A3.1)$$

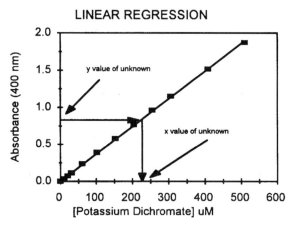

Fig. A3.2. Graphical determination of unknown concentrations using a response curve.

TABLE A3.1. Key Parameters Obtained from Regression Analyses

R^2	Goodness of fit
m	Slope of the best straight line
Std Dev$_m$	Standard deviation of the slope (SD$_m$)
b	Best y intercept when $x = 0$
Dev$_b$	Standard deviation of the y intercept (SD$_b$)

where x is the independent variable, y is the dependent variable, m is the slope, and b the y intercept where $x = 0$.

Many assays used in biochemistry are linear (at least over some limited concentration range). In such cases, it is appropriate to determine the parameters of the straight-line equation that best describe the straight-line portion of the raw data set. This is most commonly done using linear regression analysis, also called the least-squares correlation or the product-moment correlation. A regression analysis computation will yield the "goodness of fit," the slope of the best straight line[2] (m) and its standard deviation, and the best y intercept when $x = 0$ and its standard deviation (Table A3.1). The regression parameters can be used to mathematically describe the ideal straight line for the experimental data. This ideal line is often called the standard curve (see Fig. A3.3). The parameters of regression analysis can be used to determine the useful range of the assay and the limits of its use for the determination of the concentrations of samples with unknown concentrations. These regression parameters can then be used to mathematically estimate the concentration (x-axis values) of unknowns with measured y-axis values (see below).

While the regression analysis of response curve data can be computed manually (see below), such computations are complex and tedious. Today most analysts use a computer spreadsheet program's regression analysis routine to preform the regression calculations.[3]

The manually calculated computations of the various parameters in this regression output use the following equations for R^2 (correlation coefficient), constant (y intercept; b), X coefficient (slope; m), Std Err of Y Est (s_b), and Std Err of Coef (s_m). In these equations, x_{ave} and y_{ave} represent the average value of the x and y values, respectively. The y_{calc} value is the y value calcuated for a given x value using the slope and intercept terms derived from the regression analysis,

---

[2] The values of the slope and intercept obtained from a regression analysis are actually a mean slope and a mean intercept.

[3] For example, the popular spread sheets Microsoft[R] Excel, Corel[R] Quattro-Pro, and Graphical Analysis for Window's[TM] have regression analysis computation routines.

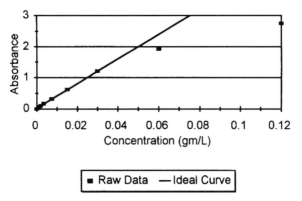

Fig. A3.3. The linear portion of a standard curve.

and n is the number of observations. The y_{calc} values are identical to the absorbance (calc) values described.

$$R^2 = \frac{\sum(x - x_{ave})(y - y_{ave})}{\sqrt{[\sum(x - x_{ave})^2][\sum(y - y_{ave})^2]}} \qquad (A3.2)$$

$$m = \frac{\sum(x - x_{ave})(y - y_{ave})}{\sum(x - x_{ave})^2} \qquad (A3.3)$$

$$b = y_{ave} - mx_{ave} \qquad (A3.4)$$

$$s_b = \sqrt{\frac{\sum(y - y_{calc})^2}{n - 2}} \qquad (A3.5)$$

$$s_m = \frac{s_b}{\sqrt{\sum(x - x_{ave})^2}} \qquad (A3.6)$$

As described, given an unknown y value, the x value can be calculated from the slope and y-intercept values of the regression analysis and:

$$X_{unk} = \frac{(y_{unk} - b_{regression})}{m_{regression}} \qquad (A3.7)$$

TABLE A3.2. Some Response Data

	x		
$[K_2Cr_2O_7]$ (g/L)	$[K_2Cr_2O_7]$ (μM)	y *A* (400 nm)	Calc y
0	0	0.000	0.006
0.003	10	0.036	0.043
0.006	20	0.075	0.080
0.009	31	0.115	0.121
0.018	61	0.235	0.231
0.030	102	0.386	0.383
0.045	153	0.578	0.572
0.060	204	0.766	0.760
0.075	255	0.960	0.949
0.090	306	1.149	1.137
0.120	408	1.517	1.514
0.150	510	1.871	1.890

Example: As an example, consider the following situation. The absorbance of a series of potassium dichromate ($K_2Cr_2O_7$) solutions was measured at 400 nm (Table A3.2). These solutions were prepared in terms of g/L (grams of $K_2Cr_2O_7$ per liter of solution). These concentration units were converted to μM using the molecular weight of $K_2Cr_2O_7$. Either set of values represents the independent variable for linear regression analysis. The absorbance represents the dependent variable. These data are plotted in Figure A3.2, where the data points are represented by symbols and the ideal curve computed from linear regression analysis results is represented by the line.

The linear regression analysis output is shown in Table A3.3.[4]
The following key questions can be answered from the regression parameters

1. How well do the experimental data fit the ideal equation $y = mx + b$?
2. What is the mean slope (m) of the best line?
3. What is the mean intercept (b) of the best line on the y axis (the signal) when the x axis (the concentration) is equal to zero?
4. What does the ideal curve look like?
5. What are the errors of m and b?
6. What is the smallest concentration that can be measured using this linear standard curve?

[4] This is an output from Corel[R] Quattro-Pro. Other spreadsheets can also be used; however, their output may look slightly different.

TABLE A3.3. Regression Parameter from Data in Table A3.2

Regression Output:	
Y intercept (b)	0.00634 AU
Std Dev of Y Est	0.009613
R Squared	0.9997
No. of observations	12
Degrees of freedom	10
Slope (m)	0.00370 AU/µM
Std Dev of slope	1.72E-05

7. What is the largest concentration that can be measured using this linear standard curve?

8. What is the useful range of this assay?

1. How well do the experimental data fit the ideal equation $y = mx + b$? The R^2 of a regression analysis gives the analyst an estimate of the quality of the fit of the data to the derived equation. An $R^2 = 1.000$ is perfect. Acceptable R^2s vary by scientific discipline. While chemical R^2s of 0.99 to 0.95 are generally considered acceptable for assays of biological systems, in most cases it is desirable to have R^2s of 0.99. Computed R^2s can be misleading. Ultimately, the analyst should look at a plot of the response curve data to determine if the response is linear with concentration. Thus, even after obtaining acceptable R^2s, one should always plot the response curve for the raw data and the ideal data to ensure that the data do fit the computed equation. If the plot of the response curve looks linear, it probably is. If it looks curved, it is probably not linear. The raw data should be randomly distributed around the ideal line. There are numerous examples where the data are not randomly distributed around the ideal line, and frequently the use of a straight-line plot is not justified, and other equations may be needed to fit the relationship of the dependent variable to the independent variables.

To demonstrate these points, consider the following three examples of x–y data. The first set of data (see Fig. A3.4)[5,6] looks like a straight line, and the linear regression analysis confirms this evaluation. These data represent a perfect straight line since the R^2 value is 1 and there is no error in the slope, X coefficient, and y intercept, constant, terms.

In contrast to Figure A3.4, the data set shown in Figure A3.5 is obviously curved. The curvature is not only evident from the data points but is made even

---

[5] It is the custom in plotting analytical data that the actual data points are represented by symbols and the computed data points are represented by lines.

[6] These regression data were generated with Corel® Quattro-Pro. With Corel® Quattro-Pro the R^2 is called the "R Squared," slope (m) is called "the x-Coefficient," its standard deviation is called "Std. Err of Coef.," the y intercept is called "Constant," and its intercept is called Std Err of Y Est."

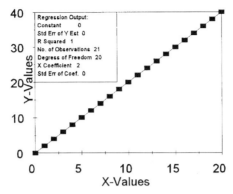

Fig. A3.4. A completely linear response curve.

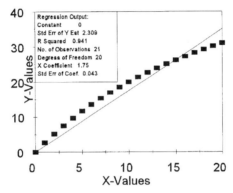

Fig. A3.5. A curved response curve.

more apparent when the regression line is superimposed on these data points. The R^2 of 0.94 regression analysis can also be interpreted to confirm the nonlinearity of the data presented in this graph. These regression analysis parameters should not be used to estimate x values from the measured y values of unknowns using a straight-line equation.

Some data sets have relatively high variations from the ideal line and still yield linear response curves. Note the plot shown in Figure A3.6. Clearly, the values in this data set are noisy. While there is scatter in these data, they do appear to represent a linear relationship when one looks at the relationship between the data points and the regression line. These response data would appear linear even though the R^2 value is less than 0.99. In this case, one could use the regression parameters for computation of unknown concentrations. However, such large variances will lead to high uncertainties in the computed results (see below), and it might be best to repeat the determination of the response curve.

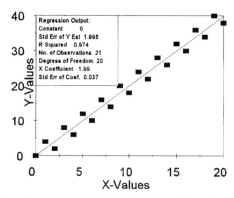

Fig. A3.6.   A linear response curve with some variation about the ideal line.

Always plot your response curve data!

In many cases of chemical analyses of biological samples, only a portion of the response curve is linear. Such nonlinearity can be due to instrument parameters such as stray light in a spectrophotometer or to chemical parameters such as insufficient reagent in a colorimetric assay. Linear regression analysis can still be applied to such data sets. However, in such cases the analyst needs to determine which part of the response curve should be used as a standard curve for the determination of unknown concentrations. The following example should make the point.

Consider the data set for a response curve shown in Table A3.4. A regression analysis results in unsatisfactory regression (see Fig. A3.7). However, in this case the analyst knew that this type of analysis tended to go nonlinear at high analyte concentrations. Thus the data sets at the higher concentrations were systematically removed from the regression analysis data set until a satisfactory R^2 was obtained. As can be seen in Figure A3.8,[7] the response curve for the lower concentrations gave a quite satisfactory set of regression parameters. Such a data set can be used for computation of unknown concentrations as long as the unknown concentrations fall within the linear range of the assay (see below).

Once an acceptable "goodness of fit" can be found for a data set, the analyst should move on to asking the other key questions as follows.

2. What is the mean slope (m) of the line?   The mean slope (m) is defined as the *X coefficient(s)* in the regression output shown in Table A3.3. The mean slope of the line indicates the ideal change of system response with a unit change in concentration. This mean slope is usually called the *sensitivity* of an assay.

3. What is the mean y intercept (b) of the line on the y axis (the signal) when the x axis is equal to zero?   The mean y intercept (b) is the theoretical

[7] The plots shown in Figure A3.8 and A3.9 were created with Graphical Analysis for Windows[R].

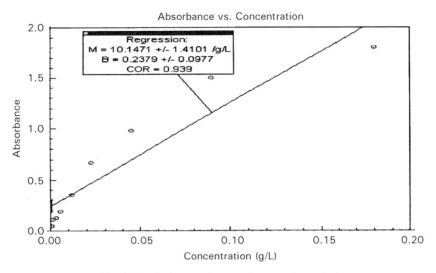

Fig. A3.7. An inappropriate use of a regression analysis.

TABLE A3.4.   Data for a Response Curve

Concentration (g/L)	ABS
0.1800	1.80
0.0900	1.50
0.045	0.980
0.0225	0.670
0.0113	0.355
0.0053	0.190
0.0030	0.130
0.0015	0.110
0.000	0.050

signal of a sample that contains no analyte. Another way of viewing the intercept value is that signal that is due to the measurement system, not the analyte.

4. What does the ideal curve look like? The ideal curve is computed from Eq. A3-1, where m and b are taken from a regression analysis with an acceptable R^2. Some spread sheets such as Graphical Analysis for Windows[TM] automatically compute this ideal line. The ideal values of y for the sample data set are also given in Table A3.2.

5. What are the errors of m and b? The standard error of the mean intercept (SD_b) is defined as the *Std Dev of Y Est* in the regression output shown in Table A3.3 and is the standard deviation ($\pm$) of b in Figure A3.8. Since b is the value of the signal with no analyte, SD_b can be viewed as the noise or variance of that no-

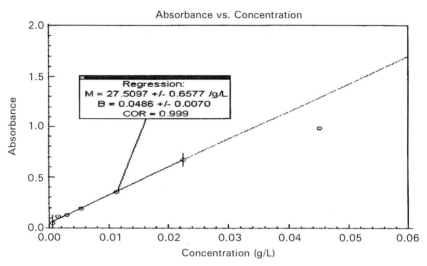

Fig. A3.8. The linear portion of the data shown in Figure A3.6 with an $R^2 \geq 0.99$. The figures of merit for this plot are shown in Table A3.5.

analyte signal. The standard error of the mean slope m (SD_m) is defined as the *Std Dev of Slope* in the regression output shown in Table A3.3 and the standard deviation of m in Figure A3.8.

6. What is the smallest concentration that can be measured using this standard curve? All analytical systems have a "no sample" background noise. It is this noise that determines the lower limit of the usefulness of the assay. The "limit of detection" (LOD) is defined as the lowest level of the analyte that can be measured.[8] The LOD has units of the x axis and is computed as:

$$LOD = 3 \ SD_b/m \tag{A3.8}$$

where SD_b is the standard deviation of the signal observed when no analyte is present in the sample and m is the sensitivity of the assay. The limit of detection for the data set in Figure A3.8 is 0.00076 g/L. The LOD is a limit value for reporting results. Even though lower levels of analyte in this particular assay would produce some signals, those signals cannot be distinguished from the noise

---

[8]Limits of detection of most detection techniques are at μmolar concentrations. A few detection techniques such as electrochemical, fluorescence, chemiluminescence can go to nmolar concentrations. Note that the practice of reporting detection limits in amounts (i.e., μmoles) is a smoke screen, since many modern assays use very small volumes without altering the limit of quantitation in concentration units.

of the assay system and thus should not be reported. Instead such values should be reported as "Not Detected, the LOD = 0.00076 g/L" (see below).

7. What is the largest concentration that can be measured using these linear regression data? The largest concentration that can be determined using the standard curve is that concentration corresponding to the last raw data point obtained from the sample with known concentration that falls on the standard curve. The largest concentration that can be determined for the analysis shown in Figure A3-8 is 0.0225 g/L. Above this concentration, an equation for a straight line should not be used to compute the concentration of an unknown concentration since the plot of the observed real data departs in a significant way from the computed ideal curve. The analyst should either graphically determine the concentration of the analyte in the unknown or dilute the sample and reanalyze it. While some might suggest that nonlinear regression analysis might be used, the inherent errors of absorption spectroscopy at high analyte concentrations usually result in limited usefulness of nonlinear regression analysis.

8. What is the useful range of this assay? The useful range of an assay is typically called the *linear range* of the assay. It runs from the LOD to the largest concentration that can be measured using this standard curve. The linear range has units of the x axis, usually concentration. In Figure A3.8, the linear range is 0.0010 to 0.0225 g/L. Beer's Law is valid only within this linear range of the response curve. In most cases the linear range for spectrophotometry is limited to 0.01 to 1.5 absorbance units; however, the linear ranges of individual spectrophotometers and assay systems can vary.

9. Figures of merit. The R^2, the sensitivity, the limit of detection, and the linear range of an assay are called the *figures of merit*. The figures of merit should be reported for any linear assay system.

10. Determination of the concentrations of unknowns. The concentration of an analyte whose absorbance was measured under the conditions used for the establishing the standard curve can be computed using Eq. A3.7. Note that when computing the mean concentrations of several replicate analyses of a single sample, it is best to compute the concentrations and then compute the mean and standard deviations of the concentrations. If a sample with an unknown concentration analyzed with the standards shown in Figure A3.8 had an absorbance of 0.555, its computed concentration would be 0.018 g/L [(0.555 − 0.0486)/27.5].

TABLE A3.5. Figures of Merit for Linear Portion of Response Curve Shown in Figure A3.9

R^2 (corr in plot)	0.999
Sensitivity (m)	27.5 ± 0.66 AU/g/L
Limit of detection	0.00076 g/L
Linear range	0.00076 to 0.0225 g/L

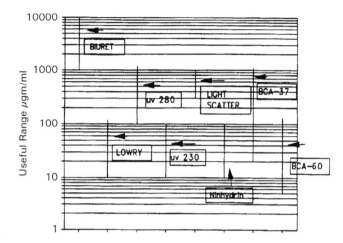

Fig. A3.9. The linear ranges of several different protein assays. Note that the limit of detection of some assays is higher than the highest point of the linear range of other assays.

Reporting Computed Concentrations

Once the analyte concentration is calculated, its value should be compared with the linear range of the assay. Only those numerical concentration values that fall within the linear range of an assay should be reported. Thus linear range can be viewed as a data filter. Those data that fall within the linear range are acceptable; those data that fall above the linear range are filtered out. Those samples with computed concentrations greater than the linear range should be reported as "for estimate purposes only, above the linear range of the assay."

Not Zero

It might seem that results that are below the limit of detection should be reported as having a zero concentration. This is not a wise action. "Zero concentration" is not a good concept for an analyst. The problem is that different assays can have different LODs. As is obvious from the data shown in Figure A3.9, different assays can have different linear ranges, and concentrations that are below the LOD of some assays are above the linear range for others. Thus for these assays, the "zero concentration" would be concentration dependent and that is foolish. Zero should not be used in reporting analytical data; rather the analyst should report not detected (ND) and give the limit of detection.

95% Confidence Limits of the Results of Computations Using Standard Curves

The standard deviation of the calculated x value ($s_{x-value}$) can be determined using the following equation:

$$s_{xvalue} = \frac{s_b}{m}\sqrt{1 + \frac{1}{n} + \frac{(y - y_{ave})^2}{m^2 \sum(x - x_{ave})^2}} \qquad (A3.8)$$

Standard Deviations of Results Computed from Standard Curve Regression Analyses

The standard deviation of the value computed for an unknown from the results from the regression analysis of a standard curve includes the variance of the results of the unknown and the variance of the results of the standard curve. Such computations are quite complex, and the reader is referred to the appropriate section of Ref. 1 of Appendix 2 for the details. However, several important points result from such computations that need to be brought to the attention of the readers of this book. First, the best precision for an assay is at the concentration that is at the midpoint of the linear curve. Second, the ninety five percent confidence limit gets larger as the concentration either increases or decreases from that midpoint. Finally, it is inherent in the computation process that the RSD of an assay is always highest at the lower concentrations of analyte (see Fig. A3.10). Note the explosion of the RSDs below the average y value on the standard curve. At markedly lower concentrations the RSDs can reach values of

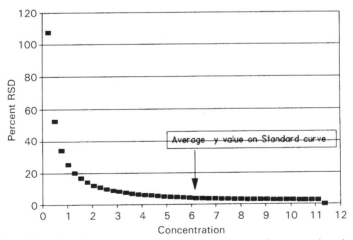

Fig. A3.10. The effect of concentration on the percent error of a computed x-value.

40, 60, or even over 100%. This explosion of an assay's RSDs is real and has been noted repeatedly in analyses at very low concentrations. If these poor precisions cause major problems, the obvious solution is to pick different assays in which the analyte concentration is equal to or higher than that value that is obtained from the average y value on the standard curve.

Appendix 4   Dilution Tables

DILUTIONS

In many cases the available stock solutions are not of the concentration needed for biochemical assays. This situation requires dilution of a stock solution to yield a new solution of a specific volume and concentration. This brief discussion is designed to remind the reader of the calculations needed to make such dilutions. First, a definition: A dilution of 1 mL to 5 mL means that you take 1 mL of solution and bring it to a final total volume of 5 mL.[1] The key concept in making dilution tables is that the *amount* of the component taken from the concentrated stock solution is equal to the *amount* of the component in the diluted solution.

$$\text{amount(stock)} = \text{amount(diluted)}$$

$$\text{amount} = \text{volume} \times \text{concentration}$$

$$\text{amount(stock)} = \text{volume(stock)} \times \text{concentration(stock)}$$

$$\text{amount(diluted)} = \text{volume(diluted)} \times \text{concentration(diluted)}$$

---

[1] This is also called a 1:5 dilution (1 to 5) since the concentration of the diluted solution is one-fifth that of the initial solution.

Linear Dilutions

Sample calculations for a series of linear dilutions are shown in Table A4.1, where the original concentration of a stock solution of compound of interest is 0.8 M and it is desired to prepare a series of samples each with a final volume of 5 mL. The following calculations were used to make the dilution series:

$$[\text{volume}_{(stock)}] \times [\text{concentration}_{(stock)}] = [\text{volume}_{(diluted)}] \times [\text{concentration}_{(diluted)}]$$

$$[\text{volume}_{(stock)}] = \frac{[\text{volume}_{(diluted)}] \times [\text{concentration}_{(diluted)}]}{[\text{concentration}_{(stock)}]}$$

$$[\text{volume}_{(stock)}] = \frac{[5mL] \times [0.2M]}{0.8M} = 1.25mL$$

Such computations are well suited for spreadsheet computations. It is a valuable exercise to compute the rest of this table using a spreadsheet to aid in your understanding of the development of dilution tables.

TABLE A4.1. Linear Dilutions

Original Stock Concentration (M)	Final sample Concentration (mL) (M)	Stock (mL)	Diluent (mL)	Final Volume (mL)
0.8	0.2	1.25	3.75	5
0.8	0.18	1.125	3.875	5
0.8	0.16	1	4	5
0.8	0.14	0.875	4.125	5
0.8	0.12	0.75	4.25	5
0.8	0.1	0.625	4.375	5
0.8	0.08	0.5	4.5	5
0.8	0.06	0.375	4.625	5
0.8	0.04	0.25	4.75	5
0.8	0.02	0.125	4.875	5
0.8	0	0	5	5

Dilution Tables for Enzyme Assays

Frequently an analyst will be given a series of stock solutions (e.g., see reagents below) and be required to make up a series of solutions with differing concentrations of some of the components. For example, one might be given the following reagents and be told to determine the rate of hydrolysis of *p*-NPP

(2 mM) using 3 mL of final solutions and six different concentrations of AP (0, 5, 10, 15, 20, and 25 µg per mL of final reaction volume; see Table A4.2).

Reagents:

Buffer- 50 mM glycine, 1 mM MgSO$_4$, 0.1 mM ZnSO$_4$, pH 10 (glycine buffer)

Stock pNPP- 20 mM p-nitrophenylphosphate (p-NPP) (biscyclohexylammonium salt) in glycine buffer

Stock AP- alkaline phosphatase (AP), 250 mg per mL glycine buffer

TABLE A4.2. Enzyme Dilution Table

Sample Number	Enzyme Concentration (µg/mL)	(mL) Stock Enz.	Concentration p-NPP (mM)	(mL) Stock p-NPP	(mL) Buffer	Total Volume (mL)
1	0	0	2	0.3	2.70	3
2	5	0.06	2	0.3	2.64	3
3	10	0.12	2	0.3	2.58	3
4	15	0.18	2	0.3	2.52	3
5	20	0.24	2	0.3	2.46	3
6	25	0.30	2	0.3	2.40	3

The dilution tables for these assays are done in a manner as for the linear dilution systems. The computations are done in steps.

1. Compute the needed volume of the p-NPP for the assays.

$$[\text{volume}_{(stock)}] = \frac{[\text{volume}_{(diluted)}] \times [\text{concentration}_{(diluted)}]}{[\text{concentration}_{(stock)}]}$$

$$[\text{volume}_{(stock)}] = \frac{[3\text{mL}] \times [2\text{mM}]}{20\text{mM}} = 0.3\text{mL Stock } p\text{NPP}$$

2. Compute the needed volume of the AP for each tube in the assays.

$$[\text{volume}_{(stock)}] = \frac{[\text{volume}_{(diluted)}] \times [\text{concentration}_{(diluted)}]}{[\text{concentration}_{(stock)}]}$$

$$[\text{volume}_{(stock)}] = \frac{[3\text{mL}] \times [\text{EnzymeConcentration}]}{\dfrac{250\mu g}{\text{mL}}}$$

for the 5 μg assay tube the volume of the enzyme stock solution

$$[volume_{(stock)}] = \frac{[3mL] \times \left[\frac{5\mu g}{mL}\right]}{\frac{250\mu g}{mL}} = 0.060mL$$

After all the dilutions have been made, the buffer is used to adjust the reaction to the final volume.

Back Calculation for Diluted Sample

If an analyst diluted a sample in the process of the assay, he/she would probably need to compute the original concentration of the analyte in the original solution before the dilution occurred. This back calculation is done as follows:

concentration of original sample

$$= \frac{(\text{concentration of diluted sample}) \times (\text{final diluted volume})}{(\text{volume of sample that was taken to be diluted})}$$

Example: Let's say the original sample was diluted by taking 3 mL of the sample and diluting it to a final volume of 20 mL. From your assay results you computed a concentration of 15 mM for the diluted sample. The original undiluted concentration is computed as follows

$$\frac{(\text{concentration of diluted sample}) \times (\text{final diluted volume})}{(\text{volume of sample that was taken to be diluted})} = \frac{15 \text{ mM} \times 20 \text{ mL}}{3 \text{ mL}}$$

$$= 100 \text{ mM}$$

Serial Dilutions

The dilutions just described are called *linear dilutions.* In some assays the dilutions are prepared as shown below. These are called *serial dilutions.*

1 mL of solution A to 100 mL final volume of solution B
1 mL of solution B to 100 mL final volume of solution C
1 mL of solution C to 100 mL final volume of solution D
1 mL of solution D to 100 mL final volume of solution E

If solution A has a concentration of 1 M, then solution B has a concentration of 0.01 M, solution C has a concentration of 0.0001 M (10^{-4} M), solution D has a concentration of 0.000001 M (10^{-6} M), and solution E has a concentration of 0.00000001 M (10^{-8} M) serial dilutions are multiplicative.

Index

Absorbance
  additivity calculations and, 51
  colorimetric reactions, product/
      reagent overlapping spectra,
      74–78
  concentration calculations, 46–49
  continuous enzyme assays,
      calculation of, 112–115
  discontinuous enzyme assays,
      calculation of, 116–119
  equilibrium colorimetric assays, 67
  linear regression analysis, 228–238
  spectrophotometric analysis:
    dilutions, 57–58
    noise errors, 56–57
  transmittance of solutions, 45
Absorption
  spectrophotometric analysis and,
      41–42
  transmittance, 44–45
Absorption photometry

additivity, 51
defined, 48
Absorption spectrum, of compounds,
    41
Absorptivity
  defined, 45–46
  spectrophotometric analysis, 50
Accuracy
  assay evaluations on basis of,
      204–205
  defined, 4
  enzyme-linked immunosorbent
      assays (ELISA), 179
  enzyme-mediated colorimetric
      reactions, validity of, 101–104
Acid-base indicators, properties of, 23
Acid-base titrations, defined, 24–26
Acid dissociation constant (K_a)
  pH buffers, 27–28
  selection criteria, 30
  weak acids, 21–22

243